# Small Houses of the Twenties

## The Sears, Roebuck 1926 House Catalog

### An Unabridged Reprint

## SEARS, ROEBUCK AND CO.

*A Joint Publication of*

THE ATHENAEUM OF PHILADELPHIA
*and*
DOVER PUBLICATIONS, INC., NEW YORK

This Athenaeum of Philadelphia/Dover edition, first published in
1991, is a republication of *Honor Bilt Modern Homes,* originally published
by Sears, Roebuck and Co., Chicago and Philadelphia, in 1926. A preface
and a publisher's note have been added.

Manufactured in the United States of America
Dover Publications, Inc.
31 East 2nd Street
Mineola, N.Y. 11501

*Library of Congress Cataloging-in-Publication Data*

Sears, Roebuck and Company.
   [Honor bilt modern homes]
   Sears, Roebuck catalog of houses, 1926 : an unabridged reprint /
Sears, Roebuck and Co.
       p.    cm.
   Reprint. Originally published: Honor bilt modern homes. Chicago :
Sears, Roebuck, 1926.
   Includes index.
   ISBN 0-486-26709-1 (pbk.)
   1. Prefabricated houses—United States—Designs and plans—
Catalogs.   2. Sears, Roebuck and Company—Catalogs.   I. Title.
NA7205.S37   1991
728'.37'0222—dc20                            91-8142
                                                   CIP

# Preface to the Athenaeum/Dover Edition

THIS REPRINT EDITION OF Sears, Roebuck's Modern Homes catalog of 1926 is one in a series of books and trade catalogs published by special agreement between The Athenaeum of Philadelphia and Dover Publications, Inc. The objective of this series is to make available to the greatest possible audience rare and often fragile documents from the extensive collections of The Athenaeum in sturdy and inexpensive editions.

The Athenaeum of Philadelphia is an independent research library with museum collections founded in 1814 to collect materials "connected with the history and antiquities of America, and the useful arts, and generally to disseminate useful knowledge." It is housed in a handsomely restored National Historic Landmark building near Independence Hall in the heart of the historic area of Philadelphia.

As the collections expanded over the past 175 years, The Athenaeum refined its objectives. Today the library concentrates on nineteenth- and early twentieth-century social and cultural history, particularly architecture and interior design, where the collections are nationally significant. The library is freely open to serious investigators, and it annually attracts thousands of readers: graduate students and senior scholars, architects, interior designers, museum curators, and private owners of historic houses.

In addition to 130,000 architectural drawings, 25,000 historic photographs, and several million manuscripts, The Athenaeum's library is particularly rich in original works on architecture, interior design, and domestic technology. In the latter area the publications of manufacturers and dealers in architectural elements and interior embellishments have been found to be particularly useful to design professionals and historic house owners who are concerned with the restoration or the recreation of period interiors. Consequently, many of the reprints in this series are drawn from this collection. The Athenaeum's holdings are particularly strong in areas such as paint colors, lighting fixtures, wallpaper, heating and kitchen equipment, plumbing, and household furniture.

The modern Athenaeum also sponsors a diverse program of lectures, chamber music concerts, and exhibitions. It publishes books that reflect the institution's collecting interests, and it administers several trusts that provide awards and research grants to recognize literary achievement and to encourage outstanding scholarship in architectural history throughout the United States. For further information, write The Athenaeum of Philadelphia, East Washington Square, Philadelphia, PA 19106-3794.

ROGER W. MOSS
*Executive Director*

# Publisher's Note

THE 1920S WERE HEADY YEARS in America economically. As the smoke of World War I cleared, the United States realized that it was the greatest economic power on earth, with an unmatched industrial and agricultural base. Consumer buying, deferred for the war years, soared in its aftermath. And "The World's Greatest Store" rode the crest of the boom.

Sears, Roebuck and Co. aspired at one time or another to be the provider of virtually every consumer product to the American people. From 1908 to 1912 Sears manufactured its own line of automobiles. For some fifteen years it sold groceries in its stores. Most ambitiously of all, from 1916 to 1933 it offered completely prefabricated houses to customers throughout the country, an offering that led quickly to the underwriting of mortgage loans (from 1911 to 1933, with one three-year hiatus) and later to actual house construction through contracting with local carpenters (from 1929 to 1934).

As early as 1909 Sears was offering plans and materials for complete homes (though without the feature of "ready-cut" lumber, labeled for easy assembly). Its generous selection of some eighty house plans remained stable in number through the years, the plans being modified for the sake of variety and to follow shifting tastes. The houses covered the broad spectrum from the truly minimal to the substantial (but not sumptuous). The present catalog depicts 72 "Honor Bilt" houses (many of them also available in reversed orientation), 8 "Standard Built" houses, 6 summer cottages, 9 garages, a hunter's cabin and an outhouse. The imposing "Glen Falls" house (pp. 42–43), at $4909, represents the top of the line, the humble "Selby" (p. 115), at $629, the bottom. (As one basis for comparison, Ford's "people's car," the Model T, sold for $290 in 1926.)

The houses' styles almost all reflect some recognizable tradition. Dutch Colonial style (see, e.g., the "Rembrandt," pp. 30–31) emphasizes the gambrel roof. English ancestry is seen in steep-pitched roofs and general verticality (see the "Barrington," p. 88). But these and other styles are far outnumbered by basic American Colonial, whose only real rival in these pages is the "bungalow," a concept Sears construes very liberally. The various colonial styles, all of which eschew nineteenth-century sawn ornamentation (brackets, vergeboards, appliqué, etc.), represent a return to a certain dignified austerity. The bungalow craze, which began around the turn of the century, was a more radical reaction oriented toward nature and rustic simplicity; by 1926 its features are only dimly visible, though the porches so integral to the bungalow concept have become almost ubiquitous.

Thus all the designs, while discarding some features from the recent past, have a certain wholesomely American lineage. The new Prairie style of Frank Lloyd Wright and others is more or less invisible, and European modernism in the form of either Art Deco or Bauhaus style is totally absent. There is nothing particularly Italian or French. And the firm apparently never attempted to create a single distinctive "Sears house."

American house building reached its pre–World War II peak in 1925, but the Modern Homes department would continue to expand through 1929, the year of the crash. Between 1919 and 1925 eleven sales offices had been opened in cities from Washington to Chicago for the sole purpose of selling these houses. By 1930 there would be 48 sales offices and a total of 350 salesmen. (Montgomery Ward, which had instituted a similar department, never exceeded twenty offices.)

The 1926 catalog—more vivid in its original color version—can be admired on many counts, not least for the uniquely persuasive prose that Sears could always deploy so skillfully. Every concern imaginable is addressed in the first twenty pages, from fundamental reasons for buying ("Give the Kiddies a Chance," "Get Close to Nature") to the makeup of the Sears Architects' Council ("including a woman adviser, who understands the requirements of the housewife") to the qualities of their wood ("King Tut's tomb, recently unearthed, revealed articles made of cypress that are still in good condition") to suggestions for paint color ("Country and suburban homes are more pleasing when decorated in light shades").

No one could have guessed that this thriving line would survive only eight more years. After nineteen years of unbroken profits, the early 1930s brought severe losses; despite many attempts at innovation, the Depression's effects could not be bucked. The home mortgages were their prime vulnerability. As "the farmer's friend," Sears had always put forth a generous image; from at least 1926 on, Sears's mortgages might finance up to 100% of a house's construction. The risks of such lending, which underwrote most of its buildings, were hard to foresee in the boom years but proved disastrous in the wake of the crash.

Thousands of readers may recognize their own homes on these pages; after all, 34,000 of the houses had been sold by 1926, and by 1934 the figure would top 100,000. These are still the houses that line our town and suburban streets; as the new millennium approaches, the reader may reflect on how slowly "the future" arrives. We hope this reissue will serve, among other things, to widen the reader's perspective to include the broad reality of American house design as distinct from the work of the leading innovators.

THE CRESCENT
Described on
Page 26

FOR
EASY
PAYMENT
PLAN
See Page 144

*Honor Bilt*

# MODERN HOMES

## Sears, Roebuck and Co.

### CHICAGO - PHILADELPHIA

# Our Liberal Easy Payment Plan

## Small Monthly Payments

Why continue to buy rent receipts when you can pay for a comfortable home of your own with your rent money? See table below how rent money piles up. Why not take advantage of our Easy Payment Plan and invest in a home of your own?

"Honor Bilt" Modern Homes are within the reach of nearly everyone. All you need is a lot and a little cash to help pay for the masonry material, which we do not furnish. We ship you all of the high grade material for a complete home, including plumbing, heating and lighting and, in some instances, loan some cash to help pay for labor and masonry material.

Our interest charge is only 6 per cent. Payments are from **$15.00 to $75.00 per month.**

**Save from $500.00 to $2,000.00 on material, labor and architecture by building an "Honor Bilt" Modern Home**

Our homes are built on honor, backed by our $100,000,000.00 Guarantee. When you buy an "Honor Bilt" Home (1) you are sure of getting high grade material throughout; (2) already cut and fitted; (3) ready for immediate construction; (4) according to easy-to-follow specifications, (5) and built for strength, durability and permanence. The best architecture and arrangement of rooms.

"Honor Bilt" Modern Homes have been erected everywhere—large cities, exclusive suburbs, small towns and on the farms. Right now there are more than 34,000 Sears-Roebuck "Honor Bilt" Homes in the U. S. A.

You can choose from over 100 beautiful designs, which are illustrated and described in this book.

### The Greatest Building Proposition!

Easy Payment Plan for "Honor Bilt" Modern Homes Described on Page 144 of This Book

*It Is the Most Liberal Home Building Plan*

598

| If Your Monthly Rent Is | You Have Spent Counting Interest at 6% | | |
|---|---|---|---|
| | In 7 Years | In 10 Years | In 15 Years |
| $25.00 | $2,518.13 | $ 3,954.20 | $ 6,982.23 |
| 35.00 | 3,525.38 | 5,535.88 | 9,775.82 |
| 40.00 | 4,029.00 | 6,326.32 | 11,172.88 |
| 50.00 | 5,036.25 | 7,908.40 | 13,965.46 |
| 60.00 | 6,043.50 | 9,490.08 | 16,768.54 |
| 70.00 | 7,050.75 | 11,071.76 | 19,551.64 |
| 80.00 | 8,058.00 | 12,652.46 | 22,345.76 |
| 90.00 | 9,065.25 | 14,234.72 | 25,138.34 |

SEARS, ROEBUCK AND CO.

# Own Your Own Home

## Long Life and Happiness

To get the full share of Good Health, Long Life and Happiness for yourself and kiddies, to get the most out of life as our Creator intended it should be, A HOME OF YOUR OWN is an absolute necessity.

It promotes happiness and contentment, for it is the most pleasant and natural way to live. It has the correct environment made up of the natural instead of the artificial.

Green grass, trees, shrubbery, flower and vegetable gardens all your own, provide a pleasant pastime, and an abundance of the things we all crave. It is the real life that leads to happiness, for you, and those you love.

Best of all, a home of your own does not cost you any more than your present mode of living. Instead of paying monthly rental, by our Easy Payment Plan you may have all these luxuries at a lower cost and, in the end, have a beautiful home instead of worthless rent receipts.

Our plan is simple. It has already enabled thousands of people to get out of the renter's class. This plan will put you in your own home and give you your independence.

On the following pages you will find over 100 designs of homes. Some of them will surely meet with your ideas of what a real home should be.

We will gladly tell you all about any house in this book and will show you how easy it is to own a home on our Easy Payment Plan. Write us. An Information Blank has been placed in the back of this book for your convenience.

Be sure to read about our Ready-Cut System on pages 10 and 11, and how this system will save about one-half of your carpenter labor.

*Information Blank on Page 141*

Save Your Rent Money

Give the Kiddies a Chance

Get Close to Nature

Have Real Friends and Neighbors

Be Independent in Old Age

**Our EASY Payment PLAN makes it POSSIBLE WHY PAY RENT?**

# Visit Our Office Nearest You And See This Exhibit

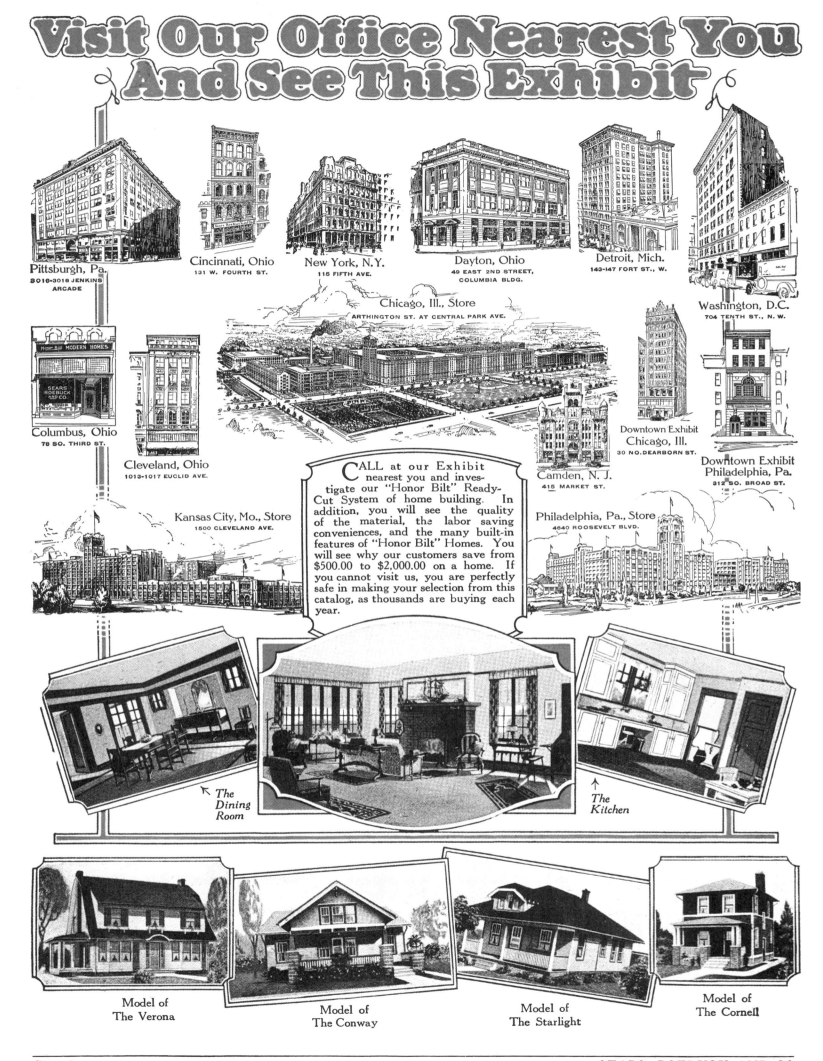

Pittsburgh, Pa.
3016-3018 JENKINS ARCADE

Cincinnati, Ohio
131 W. FOURTH ST.

New York, N.Y.
115 FIFTH AVE.

Dayton, Ohio
49 EAST 2ND STREET, COLUMBIA BLDG.

Detroit, Mich.
143-147 FORT ST., W.

Chicago, Ill., Store
ARTHINGTON ST. AT CENTRAL PARK AVE.

Washington, D.C.
704 TENTH ST., N.W.

Columbus, Ohio
78 SO. THIRD ST.

Cleveland, Ohio
1013-1017 EUCLID AVE.

Downtown Exhibit Chicago, Ill.
30 NO. DEARBORN ST.

Camden, N. J.
415 MARKET ST.

Downtown Exhibit Philadelphia, Pa.
312 SO. BROAD ST.

Kansas City, Mo., Store
1500 CLEVELAND AVE.

Philadelphia, Pa., Store
4640 ROOSEVELT BLVD.

CALL at our Exhibit nearest you and investigate our "Honor Bilt" Ready-Cut System of home building. In addition, you will see the quality of the material, the labor saving conveniences, and the many built-in features of "Honor Bilt" Homes. You will see why our customers save from $500.00 to $2,000.00 on a home. If you cannot visit us, you are perfectly safe in making your selection from this catalog, as thousands are buying each year.

↖ The Dining Room

↑ The Kitchen

Model of The Verona

Model of The Conway

Model of The Starlight

Model of The Cornell

**SEE** Page 141 for Additional Information
Page 144 for Easy Payment Offer

# REVERSED FLOOR PLANS

34,000 "Honor Bilt" Modern Homes in the U. S. A.

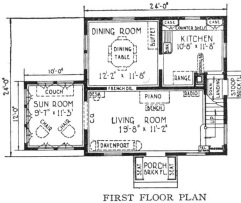

FIRST FLOOR PLAN

*Regular Floor Plan of*
**THE PURITAN HOME**
*See Pages 22 and 23*
*Second Floor Regular*
*Plan Not Shown*

THE LOT on which you plan to erect an "Honor Bilt" modern home may face such a direction that the rooms you want the most sunlight in will be on the wrong side.

You can overcome this objection by taking advantage of our reversed floor plan arrangement on many of our "Honor Bilt" Homes.

In other words, if you wish to build the house you select with the main rooms transposed, as illustrated in the plan of The Puritan, be sure to mark "REVERSED" after the name or number of the house when sending your order.

This will enable us to change some few pieces of material so they will fit when the house is built in a reversed position.

If the architectural plans permit a reverse floor plan arrangement it is mentioned in the description of the houses on the following pages.

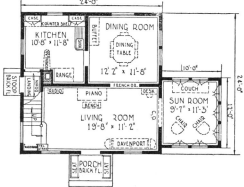

FIRST FLOOR PLAN

*Reversed Floor Plan of*
**THE PURITAN HOME**
*See Pages 22 and 23*
*Second Floor Reversed*
*Plan Not Shown*

# Over 34,000 Houses Sold

Street view of Sears, Roebuck and Co. houses at Dayton, Ohio.

Street view of Sears, Roebuck and Co. houses at Akron, Ohio.

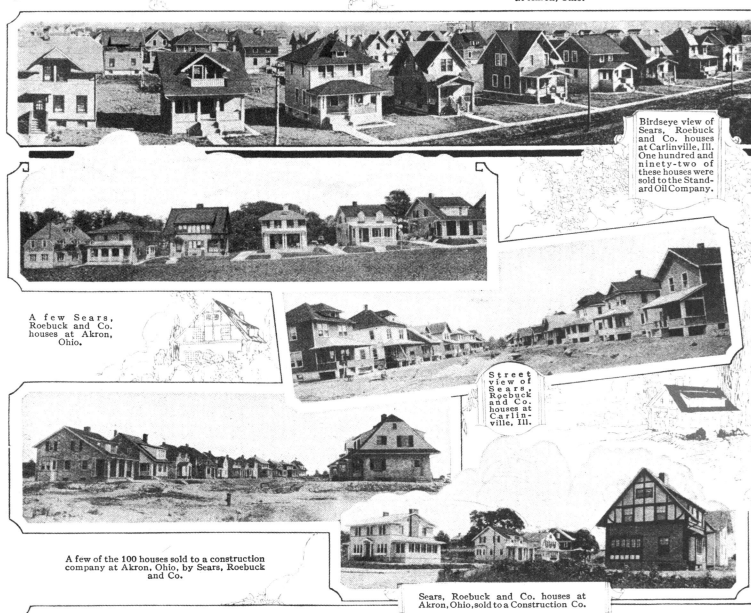

Birdseye view of Sears, Roebuck and Co. houses at Carlinville, Ill. One hundred and ninety-two of these houses were sold to the Standard Oil Company.

A few Sears, Roebuck and Co. houses at Akron, Ohio.

Street view of Sears, Roebuck and Co. houses at Carlinville, Ill.

A few of the 100 houses sold to a construction company at Akron, Ohio, by Sears, Roebuck and Co.

Sears, Roebuck and Co. houses at Akron, Ohio, sold to a Construction Co.

Sears, Roebuck and Co. houses at Wood River, Ill., sold to the Standard Oil Company.

# Every Customer Satisfied

Sears, Roebuck and Co. houses at Plymouth Meeting, Pa., sold to the American Magnesia Company.

Birdseye view of Sears, Roebuck and Co. houses at Carlinville, Ill. One hundred and ninety-two of these houses were sold to the Standard Oil Company.

## American Homes

America is facing a tremendous demand for substantial, comfortable and modern homes for its industrial and thrifty people.

Our effort in helping to meet this nation wide demand is illustrated, in part, on these two pages and, in the fact, that more than 34,000 "Honor Bilt" Modern Homes have been erected in the U. S. A. Called upon to furnish high grade materials in a hurry and to meet active competition, we immediately filled the requirements of the big operators and the large corporation with its industrial housing problem, as well as the individual home builder.

We offer in this carefully compiled Book of Homes, dependable qualities in house building materials, at prices that are as reasonable as can be made. Moreover, we place at your command the services of an organization that has been perfected to a high degree of efficiency, gained by experience of many years. See our Guarantee of Satisfaction on the back cover.

Street view of Sears, Roebuck and Co. houses at Carlinville, Ill.

A few of many Sears, Roebuck and Co. houses at Akron, Ohio.

Street view of Sears, Roebuck and Co. houses at Carlinville, Ill.

Sears, Roebuck and Co. houses at Wood River, Ill., sold to the Standard Oil Company.

# BUILT by OUR CUSTOMERS at SUBSTANTIAL SAVINGS

The voluntary words of our customers prove the wisdom in buying "Honor Bilt" Modern Homes. Here are just a few photographs and copies of parts of letters that were taken at random from our huge testimonial files. Over thirty-four thousand customers have purchased "Honor Bilt" Modern Homes.

### THE WESTLY
All material was very good and sufficient. The lumber was far superior, so carpenters said, to any that could be obtained here. We highly recommend your "Honor Bilt" houses. Hope that everyone that builds one will be as pleased as we are in cost, convenience and comfort.
ALBERT J. KEGEL,
5119 Jewett St.,
Washington, D. C.

### THE OSBORNE
We are well satisfied with our "Osborne" house. The material is as good as can be got anywhere and way above the average. I made a big saving by the use of "Honor Bilt" Ready Cut material. Our dealings with your company have been unusually satisfactory and recommend you to anyone about to build.

CLARENCE L. PARKER,
19 Olmstead Ave.,
Dearborn, Mich.

### THE AMERICUS
It is the best planned house I ever saw. Several carpenters told me it was the best material they ever used. Everything fits to a "T." I saved just $1,500.00.
JOHN HALL,
R. F. D. 2, Box 445,
Brooklyn Sta.,
Cleveland, Ohio.

### THE LANGSTON
In 1921 I bought, erected and am now living in the Americus. I have since built the Alpha and now working on a Langston and an Adeline. Your service, quality and courtesy makes me a booster for "Honor Bilt" homes.
HENRY M. JUNG,
4223 Lowry Ave.,
Norwood, Ohio.

### THE CONWAY
I built the house myself, with help only after the roof was on and the weather boarding. I estimate I saved between $1,500.00 and $2,000.00.
J. A. Paddleford,
2300 Monroe St.,
N. E.,
Washington, D. C.

### OLIVIA
I am so well pleased I am thinking of buying another.
WILLIAM BLYLY,
R. F. D. 7, Box 5,
Elkhart, Ind.

### THE ELSMORE
I figure about $1,800.00 saved by getting it Ready-Cut from Sears, Roebuck and Co.
FRED W. KROMP,
R. 1, Box 390B,
West Albany, N. Y.

# Read what they say

Every day interesting testimonials come from "Honor Bilt" Modern Home customers. They tell of the money saved, satisfaction with our quality materials, architectural plans, specifications, service and easy payment plan. There are no worries. Our guarantee protects you in every way.

**Argyle**

I have had complete satisfaction from the start. All material was of extra good grade, plenty of it, and it went together perfectly. I made a big saving in time and money.

J. O. MATTHEWS,
Xenia, Ohio.

Dear Sirs:

I want to express my appreciation for the very courteous and efficient attention that I have received from you while building my "Martha Washington" Home.

I have saved about $1,500.00, of which I credit about $1,000.00 to labor saved by the "Honor Bilt" ready cut means.

A. H. BREWOOD,
Washington, D. C.

**Alhambra**

The millwork and other materials furnished are certainly of very exceptional quality.

M. A. LANGE,
5925 Nina Ave., Norwood Park, Ill.

**Puritan**

I wish to express my appreciation of the prompt and efficient service rendered. The whole transaction has been most satisfactory. There was sufficient material and it exceeded in quality the builder's expectations.

E. E. THOMPSON,
4114 Ingomar St.,
Chevy Chase, D. C.

**Clyde**

I am well pleased with the house and with your material. My wife and I, who are nearing 60 years, built the house ourselves and we saved about $1,300.00.

W. E. O'NEIL,
715 Maple St., Wamego, Kan.

**Elsmore**

I have nothing but praise for both material and service.

JOSEPH DeHAVEN,
Glenshaw, Pa.

**Woodland**

Thank you for the efficient services rendered. The house was erected in a wonderfully short period.

LOUIS T. MACKE,
Cincinnati, Ohio.

# Skyscrapers are Ready Cut

## Honor Bilt Homes
## are Built on Skyscraper Principles
## Solid Construction with Less Labor

**Have you ever viewed a skyscraper in the course of construction? Weren't you impressed with the orderly way with which all beams were swung into place—no confusion—no cutting?**

This careful cutting and fitting of material before it is delivered on the job, whether it be steel or lumber, not only saves the cost of skilled labor, but makes for stronger and more solid construction. The use of the most modern, up to the minute machinery for the cutting and fitting, makes it possible for this work to be more accurately and economically done at the factory.

"Honor Bilt" Ready-Cut Homes are built on these same money saving, skyscraper construction principles. Here are a few interesting facts and illustrations pertaining to "Honor Bilt" Homes. Consider them very carefully.

### Real "Honor Bilt" Ready-Cut Houses

We furnish, already cut and fitted, with your "Honor Bilt" Modern Home all of the material we have found by actual experience and scientific test can be cut more economically by big power-driven machines in our factory than it can be cut by hand on the job.

Before you buy a Ready-Cut House, FIND OUT JUST HOW MUCH OF THE MATERIAL is cut and fitted. Here is what we guarantee to furnish already cut and fitted or already built for our machine made ready-cut houses:

#### Framing Already Cut

| | | |
|---|---|---|
| Basement Posts | All Treads | Brackets |
| Porch Posts | All Risers | Dental Blocks |
| Girders | Balusters | Porch and Stair |
| Wall Plates | Railing | Stringers |
| Joists | Plates | (Rough Horses) |
| Collar Beams | Rafters | Braces |
| Studding | Purlins | Trusses |

#### Trim Already Cut

| | | |
|---|---|---|
| Jambs | Stair Stringer | Stops |
| Casing | Boards Also | Seat Material |
| Stool | Housed | Apron |

#### Already Built

| | | |
|---|---|---|
| Medicine Case | China Closet | Ironing Boards |
| Cupboard | Colonnades | Breakfast Alcoves |

*Door and Window Frames Cut and Fitted*
*Doors Mortised for Locks*

We furnish our girders cut and fitted for built-up construction. Any architect or practical contractor will tell you that the built-up girder is ALMOST TWICE AS STRONG AS THE SOLID ONE.

### You Take No Chances on Quantity or Grades

Order your home direct from this book. Look over the many designs we illustrate and describe. Note our low prices. Please remember that at the prices quoted we agree to furnish ALL OF THE MATERIAL, including cellar sash and frames, with the exception of cement, brick and plaster. You will not have to buy any extras. Remember you get complete working plans, specifications and bill of material without charge (see pages 17 to 19). Our binding guarantee protects you and insures plenty of material to build your home complete. Wherever you decide to purchase a house insist that the bill includes everything we guarantee to furnish, including the painting material (we furnish three coats for outside work of "Honor Bilt" Homes), the hardware, the building paper, the sash weights, eaves trough and down spout. Then, before you hand over your hard earned money, BE SURE OF THE RESPONSIBILITY OF THE CONCERN with which you are going to deal. Be sure that you can get your money back if you want it, in case the materials do not come up to your expectations.

---

# Why Not Your Home?

**Frame Material Already Cut and Fitted**

## "Honor Bilt" Modern Homes
### Are Easy to Build

### *Every Piece Cut and Fitted Ready for Its Place*

The illustrations on this page show, better than words can tell, exactly how we make it easy for you to save a great deal on the construction cost of your new home. For actual saving, see pages 10 and 11. Here is the most difficult part of the entire undertaking reduced to such a simple proposition that you only need a hammer and nails to put up the framework of your house.

Every piece of Ready-Cut lumber required to build the complete house is numbered. As the smallest pieces are bundled and marked no time will be lost in sorting. Every number corresponds with the number shown on the plans which we furnish.

### Important Facts About Lumber

We do not handle inferior types of lumber. The lumber furnished for "Honor Bilt" Modern Homes is bright and new, fine, dry Douglas Fir or Pacific Coast Hemlock for framing, Cypress for outside finish, the wood that lasts for centuries; Oak, Birch, Douglas Fir or Yellow Pine, as specified, of selected clear grades for interior finish. If we say we give you No. 1 quality Douglas Fir, YOU ARE GOING TO GET No. 1 QUALITY. Our object in selling you an "Honor Bilt" Modern Home is to give you the kind of material that will prove to be A LITTLE BETTER THAN YOU HAVE A RIGHT TO EXPECT!

Please don't lose sight of the fact that the price is not the only consideration when you select a house. The quality must be there, too. There must be enough material to build it complete according to our plans. It must be designed with a view to comfort and convenience as well as economy. It must be well lighted, well ventilated and provision must be made for safe and satisfactory heating. All of these things must be taken into account if you are going to be permanently satisfied with your purchase, AND WE GUARANTEE TO SATISFY YOU PERFECTLY WHEN YOU BUY AN "HONOR BILT" HOME. Read our guarantee on the back cover of this book.

### Here Is Proof of Your Saving

Do you want to know what saving in labor you can make by the "Honor Bilt" Already Cut System?

A test made on August 2, 1921, showed a saving of 40 per cent or 231½ hours on a four-room house. All facts and figures with illustrations are given on pages 10 and 11.

With each house we furnish a booklet of simple directions.

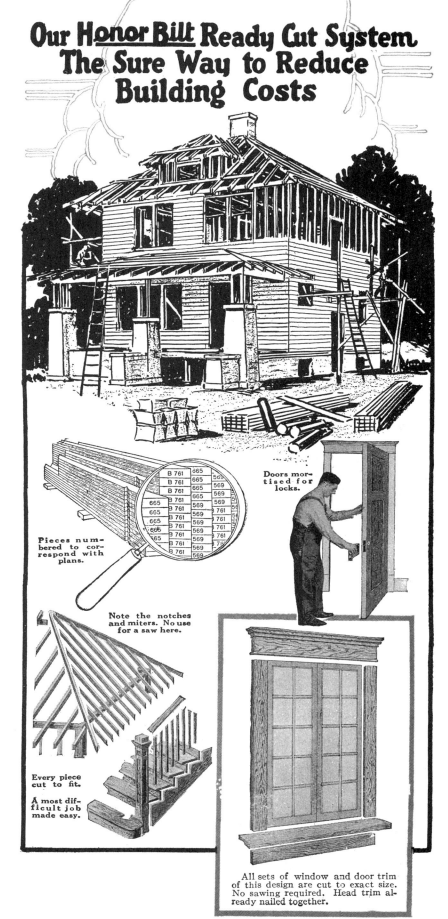

Pieces numbered to correspond with plans.

Doors mortised for locks.

Note the notches and miters. No use for a saw here.

Every piece cut to fit.

A most difficult job made easy.

All sets of window and door trim of this design are cut to exact size. No sawing required. Head trim already nailed together.

## Furnished Already Cut and Fitted

**1** STARTING. "Honor Bilt" material, all measured and sawed at our factory by modern machinery, ready to be speedily used. Watch the quick progress.

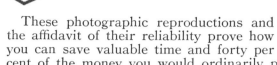

## The Honor Bilt Way

### Here is Proof

These photographic reproductions and the affidavit of their reliability prove how you can save valuable time and forty per cent of the money you would ordinarily pay for carpenter work by building an "Honor Bilt" Ready Cut Modern Home. They show the economical way and the old wasteful method of building construction.

Starting at the same time, two houses were erected side by side; yet one house was completed 231½ hours ahead of the other. When both houses were completed they were exactly alike. The material was the same. Both were substantially built. The carpenter work was as carefully done on one as the other.

**The Plain Reason:** Our "Honor Bilt" Ready Cut System is the victor in every test, every comparison. One house was built in the old fashioned, wasteful way—the other house was erected the "Honor Bilt" way, the same as "steel skyscrapers" are built.

*(Continued on the next page)*

**2** 11½ HOURS is all that was required to do all this work the "Honor Bilt" way. No time lost looking for pieces.

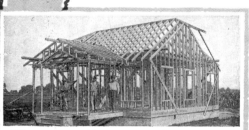

**3** 52½ HOURS carpenter labor completes the framing. No lost time—no waste material.

**Look! Read This**

**AFFIDAVIT**

STATE OF ILLINOIS }
PULASKI County, } SS.

On this—2nd—day of———August—A.D. 1921 before me L. H. Read, Notary Public, within and for the County and State Aforesaid personally came and appeared J. M. Coleman of the——City of Cairo——County of——Alexander——in the State of——Illinois——who being by me duly sworn according to law, on oath, declares that beginning July 8th, 1921 two like houses were erected as shown on the accompanying photographs, one house being furnished with framing already cut and fitted, doors mortised and inside door and window trim already cut and fitted, the material for the other house being furnished in lengths not cut or fitted. The already cut and fitted house was completely erected and finished with 352 hours carpenter labor and the house furnished with not cut or fitted lumber was completed with 583 1/2 hours carpenter labor. The accompanying photographs show the true progress of the work and waste material. Both houses being erected under the same weather conditions.

Subscribed and sworn to before me in and for the County and State aforesaid, this 2nd day of August A.D. 1921.

L. H. Read
Notary

**4** 158½ HOURS carpenter labor—house all enclosed and under roof. You can be living in your home 30 days sooner if you build the "Honor Bilt" way.

### 352 Hours Carpenter Labor

**The Honor Bilt Way**

**6** 352 HOURS carpenter labor. "Honor Bilt" house completed 231½ hours ahead of the same house built in the usual way.

**5** 281¼ HOURS carpenter labor on the "Honor Bilt" house.

# 40 per cent Carpenter Labor

## The Ordinary Way

**1** STARTING. Material on the ground. It must all be measured and cut by hand. Look at picture No. 1 on opposite page.

For one house, the lumber was supplied in random lengths just as it comes from any local lumber yard and the carpenters given a set of plans drawn in the usual way. Before they could start actual building, they had to figure their lengths, measure their boards, and cut them to proper size with hand saws. They followed the ordinary methods of building a house. They were good carpenters. They knew their business, worked fast. Yet the men on the other house were far ahead of them in a few hours.

**How the other house was built:** When the other carpenters started on the "Honor Bilt" Ready Cut House, they found their material cut to exact lengths. There was no measuring of material or cutting to do. Every piece was clearly marked to correspond with the figures on our specially drawn plans. *All guesswork was eliminated; nothing to figure, and nothing to study over. They were able to make every minute count in actually erecting the house.

The photographs tell the story. Note the mixed up pile of waste caused by cutting the material on the job on the one house; observe the absence of it on the "Honor Bilt" Ready Cut House.

Remember, these houses are typical of hundreds being built today. Go to any house that is being built the old fashioned way. Watch how slowly the work progresses. Time one of them measuring and sawing out a piece of material. Then look over these pictures and the ones on the next page again. Use your own judgment. The lumber for an "Honor Bilt" Ready Cut House is cut at the factory on large power machines. Many of the operations are automatic. This is bound to cut your labor costs.

**2** 11½ HOURS of carpenter labor on the material furnished in the ordinary way. Look at picture No. 2 on opposite page.

**3** 243 1-6 HOURS carpenter labor has been spent. Notice how the waste material is starting to pile up. Look at picture No. 3 on opposite page.

The "Honor Bilt" System in many instances has saved our customers as much as $1,000 on carpenter labor alone because it dispenses with 40 per cent (nearly one-half) of the labor cost.

*All our plans are drawn especially for our houses. Even the smallest details are so clearly shown that many people with only ordinary ability have built their own homes, and thus saved the cost of ALL Carpenter Labor.

**4** 356¼ HOURS carpenter labor. Think of the waste of material and labor.

## 583½ Hours Carpenter Labor The Ordinary Way

583½ HOURS carpenter labor. 231½ hours later than the one built the modern "Honor Bilt" way. Look at the pictures on the other page.

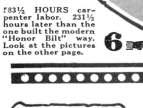

**5** 509 HOURS carpenter labor (41 days for one man) longer than was necessary.

# THE Honor Bilt SYSTEM!

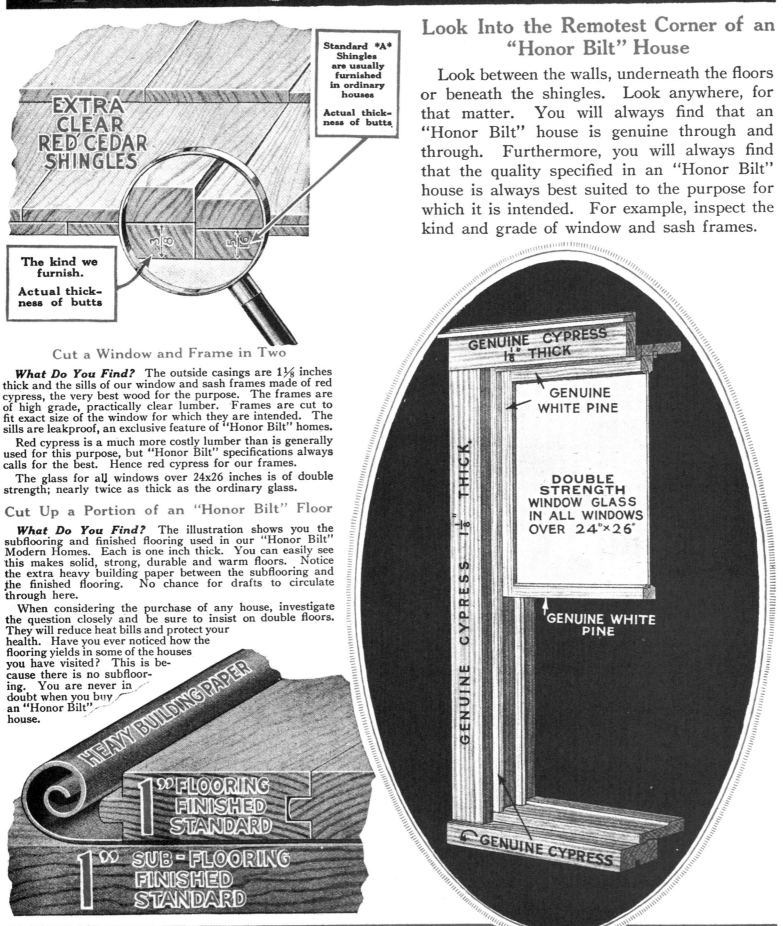

**EXTRA CLEAR RED CEDAR SHINGLES**

Standard *A* Shingles are usually furnished in ordinary houses

Actual thickness of butts

The kind we furnish.

Actual thickness of butts

## Look Into the Remotest Corner of an "Honor Bilt" House

Look between the walls, underneath the floors or beneath the shingles. Look anywhere, for that matter. You will always find that an "Honor Bilt" house is genuine through and through. Furthermore, you will always find that the quality specified in an "Honor Bilt" house is always best suited to the purpose for which it is intended. For example, inspect the kind and grade of window and sash frames.

### Cut a Window and Frame in Two

*What Do You Find?* The outside casings are 1⅛ inches thick and the sills of our window and sash frames made of red cypress, the very best wood for the purpose. The frames are of high grade, practically clear lumber. Frames are cut to fit exact size of the window for which they are intended. The sills are leakproof, an exclusive feature of "Honor Bilt" homes.

Red cypress is a much more costly lumber than is generally used for this purpose, but "Honor Bilt" specifications always calls for the best. Hence red cypress for our frames.

The glass for all windows over 24x26 inches is of double strength; nearly twice as thick as the ordinary glass.

### Cut Up a Portion of an "Honor Bilt" Floor

*What Do You Find?* The illustration shows you the subflooring and finished flooring used in our "Honor Bilt" Modern Homes. Each is one inch thick. You can easily see this makes solid, strong, durable and warm floors. Notice the extra heavy building paper between the subflooring and the finished flooring. No chance for drafts to circulate through here.

When considering the purchase of any house, investigate the question closely and be sure to insist on double floors. They will reduce heat bills and protect your health. Have you ever noticed how the flooring yields in some of the houses you have visited? This is because there is no subflooring. You are never in doubt when you buy an "Honor Bilt" house.

**HEAVY BUILDING PAPER**

**1″ FLOORING FINISHED STANDARD**

**1″ SUB-FLOORING FINISHED STANDARD**

**GENUINE CYPRESS 1⅛″ THICK**

**GENUINE WHITE PINE**

**GENUINE CYPRESS 1⅛″ THICK.**

**DOUBLE STRENGTH WINDOW GLASS IN ALL WINDOWS OVER 24″×26″**

**GENUINE WHITE PINE**

**GENUINE CYPRESS**

## Thickness of Hardwood Flooring

When we specify oak or maple flooring, we furnish it $1\frac{3}{16}$ inch thick, to be laid over the subfloor. Be sure to consider this point when comparing our prices with others.

We furnish shellac and extra durable floor varnish for our maple floors, and paste filler and extra durable floor varnish for our oak floors.

## Cut Through a Portion of the Roof

**What Do You Find?** Here are full size extra clear Red Cedar Shingles of the best quality obtainable. Their serviceable and lasting qualities are too well known to need further comment. The illustration in the upper left hand corner of the opposite page shows the extra thickness of our 5-2 Extra Clear Red Cedar Shingles we furnish when wood shingles are specified with "Honor Bilt" Homes, compared with the standard *A* grade 6-2 shingles generally furnished for most houses.

Cut and search throughout any "Honor Bilt" Modern Home; you will find every detail in every section represents the choicest material.

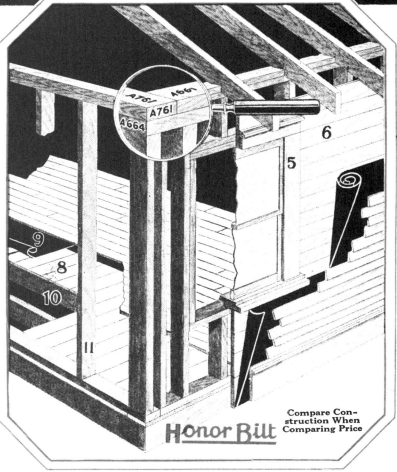

Compare Construction When Comparing Price

Honor Bilt

*"Honor Bilt" Modern Homes are illustrated and described on pages 1 to 112, inclusive*

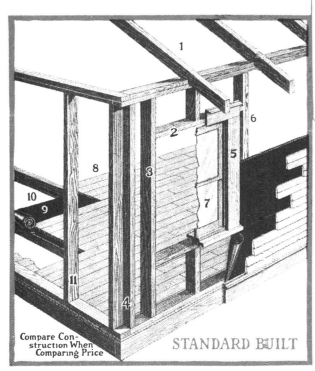

Compare Construction When Comparing Price

STANDARD BUILT

### Standard Built Construction
#### (See picture above)

1—Rafters, 2x4 inches, $22\frac{3}{8}$ INCHES APART.
2—SINGLE PLATES over doors and windows.
3—SINGLE STUDDINGS at sides of doors and windows.
4—TWO STUDS at corners.
5—Outside casing $\frac{3}{4}$ INCH THICK.
6—NO wood sheathing.
7—All glass, SINGLE STRENGTH.
8—NO SUB-FLOOR.
9—Tarred felt under floors and siding.
10—Joists, 2x8, are placed $22\frac{3}{8}$ INCHES APART.
11—Studdings, 2x4 inches, $14\frac{3}{8}$ INCHES APART.
12—Star "A" 6-2 Red Cedar Shingles for roof.
13—All outside paint, two coats.

*Standard Built Homes are illustrated and described on pages 113 to 120, inclusive.*

## "Honor Bilt" Is the Better Home for You

**Here Are the Reasons:**

An "HONOR BILT" home means a home of guaranteed quality. It means the best in quality of workmanship and in quality of material—also architectural and free plan service (see pages 17 to 19). Judge for yourself by examining the two illustrations on this page. See the difference between Standard Built construction and "HONOR BILT" construction.

Naturally, a Standard Built house will cost less than an "HONOR BILT" house of the same size. But the thirteen reasons clearly explain why the "HONOR BILT" is well worth the low price we charge.

### "Honor Bilt" Construction Illustrated Above

1—Rafters, 2x6 or 2x4 inches (larger where needed), $14\frac{3}{8}$ INCHES APART.
2—DOUBLE PLATES over doors and windows.
3—DOUBLE STUDDINGS at sides of doors and windows.
4—THREE STUDS at corners.
5—Outside casing, $1\frac{1}{8}$ INCHES THICK.
6—High grade WOOD SHEATHING, $1\frac{3}{16}$ inch thick.
7—All glass over 24x26 inches is HIGH QUALITY DOUBLE STRENGTH.
8-9—DOUBLE FLOORS WITH HEAVY BUILDING PAPER between the subfloor and finished floor.
10—2x8-inch joists, or 2x10 where needed, $14\frac{3}{8}$ IN. APART.
11—Studdings, 2x4 inches, $14\frac{3}{8}$ INCHES APART.
12—Best Grade of clear Cedar Shingles, Oriental Asphalt Shingles or Oriental Slate Surfaced Roll Roofing, guaranteed for seventeen years, as specified.
13—All outside paint, three coats of guaranteed paint, shingle stain (when shingles are used as siding), two brush coats.

# Honor Bilt Homes Factory

THESE are reproductions of actual photographs taken for the purpose of giving our customers a better idea of the tremendous stocks carried at our lumber mills and millwork factories IN READINESS TO SHIP AT ONCE. Here you will find a larger stock of building material than will be found carried by any retail dealer.

*ABOVE*—MODERN HOMES DIVISION

As shown in pictures, our millwork is carefully stored in dampproof, dirtproof buildings, assuring our customers of receiving only bright, clean stock. Here you will find about 90,000 Doors, over 100,000 Windows, millions of feet of hard and soft wood Moldings and Columns, Newels, Stair and Porch Material; also Flooring in equally large quantities.

*ABOVE*

A birds-eye view of our 17-Acre Sash and Door Factory in Ohio.

*LEFT*—DOOR WARE-HOUSE

*ABOVE*—ORDER PRICING DIVISION

This book of Modern Homes shows a great variety of homes suitable for every section of the country: East, west, north and south. All our designs have proved successful. We guarantee our plans to be correct. Select your house direct from this book and send us your order today! Otherwise write us for full information about the house you are interested in. Our architects are at your service.

*ABOVE*—BUILT-UP COLUMN WAREHOUSE

Pages 17, 18 and 19 portray just a glimpse of our architectural department facilities. We exercise every care in originating house plans for all requirements.

# Facilities and Offices

EVERY day hundreds of orders for materials and requests for plans and estimates come from all parts of the country. Efficient workers give every kind of service. We have the largest organization of its kind in the world. Our architects, draftsmen and expert women advisers have mastered home building problems.

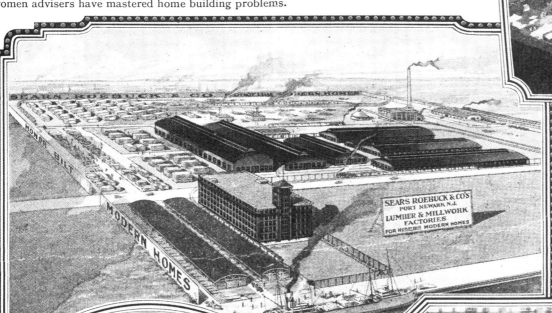

*ABOVE*—MODERN HOMES DIVISION

*LEFT*—A birdseye view of our huge Newark, N. J., Lumber and Millwork Plant. It covers a total of 40 acres, connecting with railroad terminals and ocean going freighters. The plant consists of fourteen manufacturing buildings and has the latest and most efficient modern machinery. Good service is assured, because the plant has a capacity of 100,000,000 feet of lumber per annum.

*LEFT*—MOLDING WAREHOUSE

*RIGHT*—MODERN HOMES DIVISION

These modern facilities and mammoth stocks make it possible for us to furnish you EXTRA GOOD QUALITY and EXTRA QUICK SERVICE, while our prices enable you to make A SUBSTANTIAL SAVING.

*BELOW*—A BIRDSEYE VIEW OF OUR BIG 40-ACRE LUMBER PLANT IN ILLINOIS

*ABOVE*—WINDOW WAREHOUSE

If you plan erecting a house not described in this book, send us a full list of the material you need and get our free estimate. We are in a position to quote low prices on lumber, millwork, roofing, hardware, plumbing goods, heating plant, electric lighting fixtures, paint and wall paper in any quantity.

# Better Building Material

One of ten 800-foot loading platforms at our big lumber yards in Illinois, insuring Service and Quality. Already Cut lumber shipped from this stock direct to you.

These solidly constructed rainproof warehouses protect our millwork lumber until it is required in this factory. Our high quality millwork shipped from this clean, bright and dry stock direct to you.

A trip through our Sash and Door Plant, which covers nearly twenty acres, reveals the secret of our low prices, our excellent service and the high quality of this factory's products. Here in these lumber storage warehouses you would find from twenty to thirty million feet of the choicest lumber for manufacturing high grade millwork and Ready-Cut houses.

## Millions of Feet

In the storage warehouses illustrated above you would find genuine White Pine in various thicknesses and lengths, some of the boards measuring as much as 24 inches in width, that was grown in the mountains of California. You would see vast quantities of high grade Oak, grown in the forests of Kentucky; millions of feet of choicest Douglas Fir or Pacific Coast Hemlock from the states of Washington and Oregon, and Birch from Northern Wisconsin. In our sheds would be found millions of feet of the famous Louisiana Red Cypress, which is used extensively by us for exterior construction, such as outside casings, siding, etc., a wood that is generally too high priced for this purpose. Of all the lumber mentioned, Cypress is the most weather resisting of them all. King Tut's tomb, recently unearthed, revealed articles made of cypress that are still in good condition, notwithstanding the thousands of years they had been buried under the ground. Cypress is justly called "The Wood Eternal," and, on account of its wonderful weather resisting qualities, is specified very liberally by us in all places where material is exposed to the elements.

## Scientific Drying

The greatest pains are taken to buy the best lumber, regardless of the distance the forest may be from our factory. We also use special care to prevent deterioration, warping or season checking. Lumber of every kind that enters into a millwork product must be well air dried before being placed in our storage warehouses, and then directly before using must pass through kilns that eliminate excess moisture. Kiln drying today is a scientific process. If lumber is kiln dried too much, doors and windows are liable to expand when in use. If kiln dried too little, heat in a building is liable to cause shrinkage. Therefore, all lumber before it enters into our finished products is subjected to a laboratory test to determine the exact moisture content, before the material can be used. Hence, the millwork used in our Modern Homes is entirely dependable and less liable to shrinkage, expansion or warping than millwork made in the customary way, or according to old processes.

## Our Great Transfer System

All lumber carried in the Storage Warehouse (see illustration at top of this page) is carefully piled on movable trucks at the time the lumber is graded out of the incoming cars. Electric power transfers every load to its proper place in these vast storage warehouses. There the lumber remains until thoroughly seasoned and air dried. When needed, it is transferred onto the central rails through the dry kilns, and into the cutting rooms, without ever having been removed from the original trucks or touched by human hands. To our knowledge there is no other arrangement that is as productive of high quality and low cost of handling as the one in use at our Millwork Factory.

## Good Lumber Makes Good Millwork

As the result of the care we have taken in the warehousing of our lumber, and manufacturing it into doors, windows, moldings, etc., we daily receive unsolicited letters from contractors and buyers of "HONOR BILT" houses. They tell us that the millwork we handle is of a much better grade and quality than they can buy in their local market. Here is what one contractor says about our millwork:

"—The lumber was an excellent grade and I had people tell me that it had been a long time since they saw where the window frames, trim and siding were all Cypress. Your construction is very good, as your doors, windows, the corners and headers are all double, making strong houses—also to say that your cuts are good fits—better than could be done by hand."

Elmer M. Blume, Des Plaines, Ill.

In our files are thousands of letters similar to the above. Such recommendations prove that our millwork is far better than the kind that is generally used in houses.

## Even Framing Lumber Is Protected

In the lower illustration we show one of our five 800-foot double deck lumber warehouses at our big Lumber Plants where forty acres are devoted to the sorting, manufacturing and shipping of lumber for "HONOR BILT" Modern Homes. We maintain the same system in our Eastern plant. Compare these facilities with those of the average lumber yard in your vicinity. Framing lumber, sheathing, siding, etc., in retail yards, is usually piled out in the open, and not sheltered from rain, sun, snow or dirt. Here you can see that all of our lumber, even including that which is used for framing, such as studding, rafters, etc., is kept clean and dry, free from rot or discoloration. Result: "HONOR BILT" houses are of better material and have a longer life than frame houses built in the ordinary way.

## Lumber From Virgin Forests

The framing lumber used in the construction of Ready-Cut "Honor Bilt" Modern Homes finds its origin in the virgin forests of Washington and Oregon, where the best close grained Douglas Fir and Pacific Coast Hemlock are grown. Compare this high grade framing material with Tamarack, Spruce and Huron Pine that is being used in many districts for this purpose. You will then readily understand why "Honor Bilt" Modern Homes are better than other frame constructed buildings.

## The Man Who Knows

To prove that the man who knows appreciates quality when he sees it, we are publishing below a letter from a contractor who has just completed one of our Ready-Cut "HONOR BILT" Modern Homes:

"—I have personally built about twenty-five of your houses and am well acquainted with your material and construction. The fact that I am wanting one of your houses now for myself testifies sufficiently as to my opinion of your material, construction and treatment."

E. G. Eckenrode, 23 No. 13th St., Kenmore, Ohio.

Limited space does not permit our publishing more letters of this kind, but we have thousands of them. Write us asking for names and addresses of customers residing in your state or county.

# Architectural Service

## What's it worth to you?

We make no charge for this valuable Service when you make your selection from this book

The usual charge is—from $100.00 to $1000.00 for what we give you FREE

ARCHITECT

The scale of prices charged by architectural associations vary according to agreement, but the following is typical:

Dwellings costing less than $10,000.00 — 10 per cent. Dwellings costing more than $10,000.00 from 6 per cent to 10 per cent on the completed cost.

"Honor Bilt" Modern Homes are guaranteed to be architecturally correct. You are absolutely protected against mistakes of any kind chargeable to us. They measure up to the highest standard of well built frame construction and materials; and in addition, you gain by the economical advantage of our Ready Cut system.

# 5 Reasons Why
## this Service will serve you best
## It Saves You—

**1 Saves you**

### All Architects' Fees

Good architectural service, when engaged in the customary way, costs from 6 to 10 per cent of the total cost of the house. Manufacturing enormous quantities of material in our own factories, and selling direct to you, lowers our cost of doing business to a point where we can afford to give you this valuable service without an additional charge.

**2 Saves you**

### From Experimenting with Untried Plans

There is considerable risk and disappointment when building with unproven plans. "HONOR BILT" plans are guaranteed free from architectural errors. In event of there being errors of any kind, Sears, Roebuck and Co. will stand the entire expense. Where else can you match such a dependable service, so free from risk or loss?

**3 Saves you**

### From Buying Extras and Surplus Materials

An "HONOR BILT" customer never need worry about the dreaded extras. They know before they start what the cost is going to be. If by chance the material proves insufficient, or of the wrong kind, Sears, Roebuck and Co. pay for the extra material. Did you ever hear of anyone else who did?

**4 Saves you**

### From Poor Workmanship and Faulty Construction

"HONOR BILT" Ready Cut Houses are cut to a mathematical correctness—every foot of framing material is cut by machine—standardized on the same principle as a good automobile. This prevents "butchered carpentry," and results in a more rigid construction than the ordinary hand framed job.

**5 Saves you**

### On Cost of Erection

Many wise contractors prefer "HONOR BILT" Ready Cut Houses in preference to not ready cut. There's a reason. It saves time and money, and makes satisfied customers. No better proof of our Ready Cut system can be found than the recommendation of the man who knows. Don't fail to read all about the labor and time saved by our "HONOR BILT" Ready Cut system—see pages 10 and 11.

## See
## Following Pages for Our Complete Service

# This Service Is FREE
## to Buyers of Honor Bilt Homes

**ELEVATIONS**
Front-Rear-Right-Left

**FLOOR PLANS**
First Floor-Second Floor-Basement

**ERECTION PLANS**
Locating Ready Cut Lumber

Instructions for **Installing Modern Plumbing Systems**

**Plumbing Installation**

**HOW TO BUILD** THE READY CUT WAY
FOR THE MAN WHO BUILDS HIS OWN HOME
**HOW TO BUILD READY CUT HOUSES**

I **Contractor's Proposal Forms** & II **Construction Contract Forms With Complete Specifications**

**Certificate OF Guarantee**
SEARS ROEBUCK AND CO.

**A GUARANTEE BACKED BY $100,000,000**

**WIRING HOUSES** FOR THE ELECTRIC LIGHT BY NORMAN H. SCHNEIDER

**CONTRACT Between Owner and Contractor**

**ELECTRIC LIGHTING DIRECTIONS**

**Bill of Materials** for Honor Bilt Modern Home for _____

**BILL OF MATERIALS**

## Architects' Council

Where else is a house plan given the careful study and the benefit of such experience? This Council consists of a Chief Architect, and a corps of able assistants, including a woman adviser, who understands the requirements of the housewife.

**DETAIL SECTIONS of Construction**

**DETAILS OF INTERIOR Trim and Cabinets**

Instructions for Installing and Operating Our
**Hercules Warm Air Pipe Furnace**
Sears, Roebuck and Co.

**HERCULES HEATING SYSTEMS**
Sears, Roebuck and Co.

Specifications For "Honor Bilt" Modern Homes
We Guarantee
Sears, Roebuck and Co.

**HEATING INSTALLATION**

## Our Free Service

OUR FREE SERVICE gives you all the architectural service, items 1 to 8, commonly furnished by architects and, in addition, complete Ready Cut plans, bills of material, directions for installation for all trades, etc., items from 9 to 15.

¼ Inch Scale Drawings or Blue Prints, and Specifications, such as:

1. Front Elevation.
2. Rear Elevation.
3. Left Elevation.
4. Right Elevation.
5. Floor Plans.
6. Basement Plan.
7. Detail of Interior.
8. Specifications.

*We furnish all the plans and specifications described to the left, and all these besides:*

9. Complete Erection Plans, showing all the framing material cut to exact measure (see pages 8 and 9), marked with numbers that show where every piece belongs, and which correspond with the numbers marked on all framing material.
10. Complete Itemized Bill of Material of everything that we furnish needed to build the house.
11. Contractor's Proposal Form.
12. Construction Contract legally correct.
13. Special Instructions on how to build with no loss of time.
14. Directions for Installation of:
    (a) Steam or Hot Water Heating.
    (b) Warm Air Heating.
    (c) Plumbing.
    (d) Electric Wiring.
15. **A Certificate of Guarantee that takes all the worries out of building.**

Moreover, "HONOR BILT" Architectural Service is the only service where the architect really guarantees perfect plans, and one price to cover everything —there are NO EXTRAS when you build the "HONOR BILT" way.

"HONOR BILT" Architectural Service has made standardization possible —a very big saving in itself. Due to our enormous volume, our cabinets, bookcases, stairs, cupboards, disappearing ironing boards, tile sinks, breakfast alcoves, buffets, mantels, closets, and other special features, are stock items, made in large quantities, the same as our doors, windows, trim, etc., giving you the special features that, if built to order, would greatly increase the cost of the house.

"HONOR BILT" construction is the better construction, otherwise it could not pass the rigid codes of the larger cities.

Can you afford to pass up this wonderful Architectural Service, when it is offered you WITHOUT ONE PENNY OF COST, if you are a buyer of an "HONOR BILT" Home?

### EASY TO BUILD

C. E. Anderson,
Deerfield, Ill.
"I figure I saved over $900.00 on the 'Argyle,' and it went up in fine shape. Anyone handy, with a little help, could easily build his own house by following your easy working plans."

W. H. Money,
Camp, Ohio.
"The framing all fit perfectly and agreed with your plans, which were easily understood. By building the house myself I figure I saved $1,000.00 and the lumber was all first class."

*Our Plans Are Simple and Easy to Follow*

Even customers with no previous building experience and without skilled help have successfully built "Honor Bilt" Modern Homes.

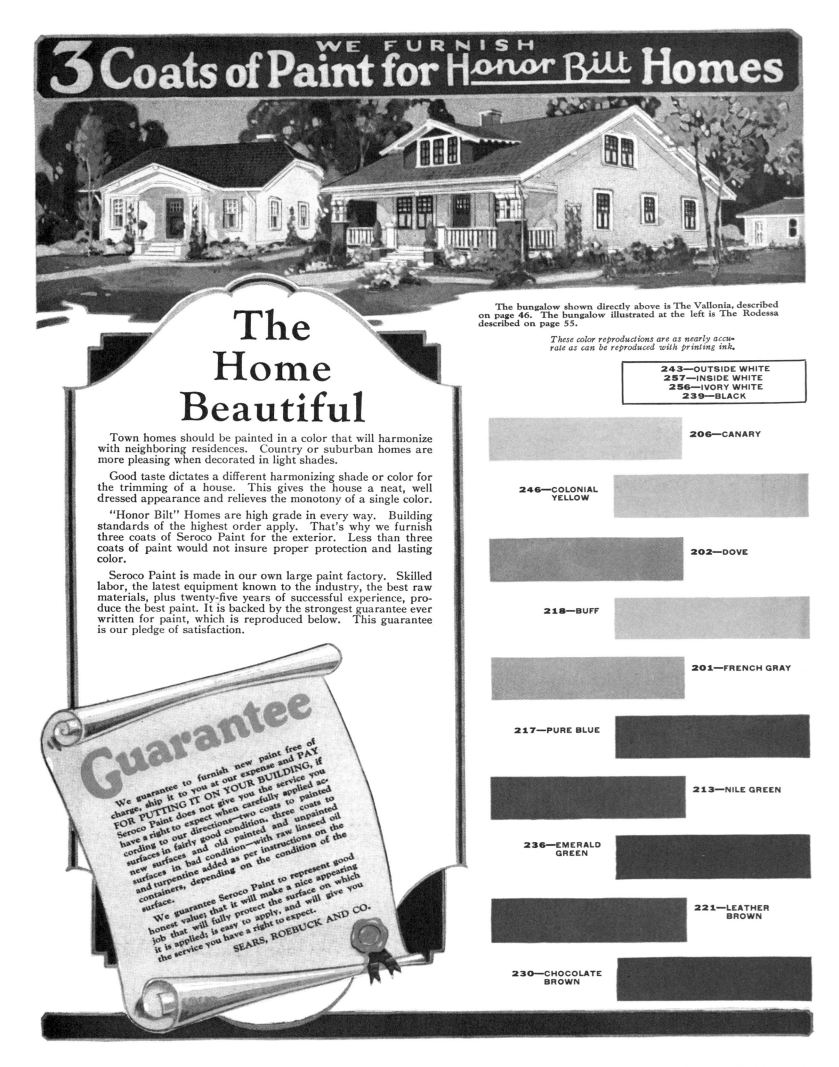

# 3 Coats of Paint for Honor Bilt Homes

WE FURNISH

The bungalow shown directly above is The Vallonia, described on page 46. The bungalow illustrated at the left is The Rodessa described on page 55.

*These color reproductions are as nearly accurate as can be reproduced with printing ink.*

# The Home Beautiful

Town homes should be painted in a color that will harmonize with neighboring residences. Country or suburban homes are more pleasing when decorated in light shades.

Good taste dictates a different harmonizing shade or color for the trimming of a house. This gives the house a neat, well dressed appearance and relieves the monotony of a single color.

"Honor Bilt" Homes are high grade in every way. Building standards of the highest order apply. That's why we furnish three coats of Seroco Paint for the exterior. Less than three coats of paint would not insure proper protection and lasting color.

Seroco Paint is made in our own large paint factory. Skilled labor, the latest equipment known to the industry, the best raw materials, plus twenty-five years of successful experience, produce the best paint. It is backed by the strongest guarantee ever written for paint, which is reproduced below. This guarantee is our pledge of satisfaction.

## Guarantee

We guarantee to furnish new paint free of charge, ship it to you at our expense and PAY FOR PUTTING IT ON YOUR BUILDING, if Seroco Paint does not give you the service you have a right to expect when carefully applied according to our directions—two coats to painted surfaces in fairly good condition, three coats to new surfaces and unpainted and painted surfaces in bad condition—with raw linseed oil and turpentine added as per instructions on the containers, depending on the condition of the surface.

We guarantee Seroco Paint to represent good honest value; that it will make a nice appearing job that will fully protect the surface on which it is applied; is easy to apply, and will give you the service you have a right to expect.

SEARS, ROEBUCK AND CO.

243—OUTSIDE WHITE
257—INSIDE WHITE
256—IVORY WHITE
239—BLACK

206—CANARY

246—COLONIAL YELLOW

202—DOVE

218—BUFF

201—FRENCH GRAY

217—PURE BLUE

213—NILE GREEN

236—EMERALD GREEN

221—LEATHER BROWN

230—CHOCOLATE BROWN

WHERE can you find a more imposing and dignified study in modern colonial architecture?

The wide expanse of pure white, contrasted with green shutters, the red brick chimneys and the green roof, surely will appeal even to the most esthetic. Observe the stately pilasters at the corners; the dentals beneath the eaves and in the gables; the balance of design afforded by the sun parlor at the left and the dining porch on the right. Observe the stately colonial porch, in harmony with the rest.

### FIRST FLOOR

**The Reception Hall.** On entering, you pass through the colonial entrance, a quaint colonial door with side lights on either side, giving light and cheer to a reception hall nearly 10 feet wide, that connects through French doors to the living room on the left and the dining room on the right. Here, too, is a beautiful colonial stairway, a clothes closet and a door leading to the first floor lavatory. Both openings have mirror doors, which appear to add length to the hall. There is room for a grandfather clock. On each side of the entrance wardrobes are provided.

**The Living Room.** From the reception hall you pass through the French doors on the left and enter the large living room, nearly 13 feet by 23 feet 3 inches, which has a real colonial fireplace.

**The Sun Room.** Directly adjoining the living room is a large sun room with French windows on three sides.

**The Dining Room and Dining Porch.** Retracing back through the hall, and to the right, you enter the dining room and, directly to the right, you enter a dining porch which has French windows on two sides.

**This house can be built on a lot 55 feet wide**

**Honor Bilt**

## The Lexington
### No. P13045 "Already Cut" and Fitted
### $4,365.00

**The Kitchen.** A door from the dining room leads to the kitchen and a passage is provided that connects the dining porch with the kitchen and a large pantry which has ample shelving and space for refrigerator that is iced from the outside. The double windows supply an abundance of light and fresh air. The kitchen cabinet is furnished with bread board, dough board, flour bins and work table. The arrangement keeps the iceman and the milkman from intruding, to say nothing of tracking the floors. A convenient grade entrance leads outside and also to the basement.

### SECOND FLOOR

**The Bedrooms.** Here are three fair size bedrooms and maid's room with good size closets with shelves in each room. A linen closet is located in the second floor hall and there is a stairway leading to a large attic. A bathroom is at the end of hall, convenient to all bedrooms. Two French windows are located on second floor landing, giving plenty of light to the stairway and second floor hall. Doors lead from bedrooms to balcony over sun room.

**Basement.** Room for furnace, laundry and storage.

**Height of Ceilings.** Basement is 7 feet from floor to joists. Main floor rooms are 9 feet from floor to ceiling. Second floor rooms are 8 feet 2 inches from floor to ceiling.

### What Our Price Includes

At the price quoted we will furnish all the material to build this nine-room colonial house, consisting of:
Lumber; Lath;
Oriental Asphalt Shingles, 17-Year Guarantee;
Siding, Clear Cypress;
Framing Lumber; No. 1 Douglas Fir or Pacific Coast Hemlock;
Flooring, First Floor, Clear Maple; Second Floor, Bathroom Floor, Clear Maple; Balance of Rooms and Hall, Clear Douglas Fir or Pacific Coast Hemlock;
Porch Ceiling, Clear Grade Douglas Fir or Pacific Coast Hemlock;
Finishing Lumber;
High Grade Millwork (see pages 110 and 111);
Interior Doors, Birch for Dining Room, Living Room and Hall; All Other Doors Two Cross Panel Design Douglas Fir;
Trim, Birch for Living Room, Dining Room, Hall and Stairs; Balance of Rooms, Beautiful Grain Douglas Fir or Yellow Pine;
Kitchen Cabinet; Medicine Case;
Colonial Brick Mantel;
Windows of California Clear White Pine;
Building Paper; Sash Weights;
40-Pound Building Paper;
Eaves Trough and Down Spout;
Chicago Design Hardware (see page 132);
Paint for Three Coats Outside;
White Enamel and Mahogany Stain for Trim and Doors of the Living Room, Dining Room, Hall and Stairs on First Floor;
Shellac and Varnish for All Interior Trim and Doors except in Living Room, Dining Room and Hall on First Floor;
Shellac and Varnish for Maple Floor.
Complete Plans and Specifications.
Built on a concrete and brick foundation and excavated under entire house.
We guarantee enough material to build this house. Price does not include cement, brick or plaster.
See descriptions of "Honor Bilt" Houses on pages 12 and 13.

### OPTIONS

*Sheet Plaster and Plaster Finish, to take the place of wood lath, $311.00 extra. See page 109.*
*Tile Floor in bathroom and toilet, instead of maple, $26.00 extra.*
*Storm Doors and Windows, $173.00 extra.*
*Screen Doors and Windows, galvanized wire, $98.00 extra.*

For prices of Plumbing, Heating, Wiring, Electric Fixtures and Shades see pages 130 and 131.

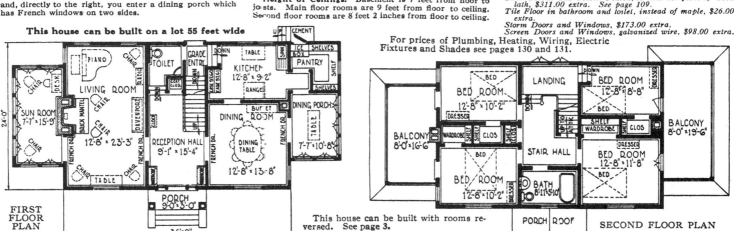

FIRST FLOOR PLAN

This house can be built with rooms reversed. See page 3.

SECOND FLOOR PLAN

*For Our Easy Payment Plan See Page 144*

THE PURITAN is the most modern type of Dutch Colonial architecture. Painted pure white with contrasting green shutters, and the red or green roof with Colonial red brick chimney, it is an architectural masterpiece. Where will you find a more inviting entrance than this quaint Colonial doorway with Colonial hood, which can be ornamented by the Colonial benches on either side of the doorway? The Puritan has a larger living room, a more spacious dining room and a more roomy kitchen than is found in houses at nearly twice its price. The Puritan, with a foundation of only 24 feet by 24 feet, has just as much real living room space as many houses with dimensions of 34 feet by 34 feet.

### FIRST FLOOR

**The Living Room.** This large living room with two full size windows providing light and ventilation, measures over 19 feet 8 inches by 11 feet 2 inches. To the right an open stairway to the second floor carries out the true Colonial charm. A clothes closet is on the stair landing within a few steps of the front door. Opposite the front entrance a large wall space affords room for a grand or upright piano. Phone shelf provided near stair landing.

#### Can Be Built on 33-Foot Lot
Can be built with the rooms reversed. See page 3.

**Honor Bilt**

## The Puritan
### "Already Cut" and Fitted
**P3190A with Sun Room, $2,475.00**
**P3190B without Sun Room, $2,215.00**

**The Sun Room.** Size, 9 feet 7 inches by 11 feet 3 inches. A cased opening leads into the sun room from the living room if P3190A is ordered. Seven pairs of French windows flood this room with sunshine. Tastily furnished and draped, it can be made the chosen spot in winter or summer.

**Dining Room.** Size, 12 feet 2 inches by 11 feet 8 inches. French doors lead from living room to dining room. A wide casement window is on the rear wall just over the space for the buffet. Two regular size windows give plenty of light from the sides. A swinging door opens into kitchen.

**The Kitchen.** Size, 10 feet 8 inches by 11 feet 8 inches. The range and sink are handy as we enter. A kitchen case makes the work of food preparation easy. A long work table with five drawers is directly underneath the window. Ample space for a refrigerator. There is a built-in ironing board. The grade entrance helps to keep the kitchen clean and warm. It also makes it easy to reach the basement from the yard or kitchen.

### SECOND FLOOR

**The Bedrooms.** A spacious hall connects all rooms. Facing the stairway is the bathroom. Each of the three bedrooms has two or three windows, and a clothes closet. A linen closet is in the hall.

**Basement.** Room for furnace, laundry and storage.

**Height of Ceilings.** Basement, 7 feet from floor to joists, with concrete floor. First floor, 9 feet from floor to ceiling. Second floor, 8 feet 2 inches from floor to ceiling.

### OPTIONS

*Sheet Plaster and Finish, to take the place of wood lath for P3190A, $206.00 extra; for P3190B, $189.00 extra. See page 109.*
*Oriental Asphalt Shingles, guaranteed 17 years, for roof instead of wood shingles, $38.00 extra.*
*Seats for Porch, $29.00 per pair extra.*
*Storm Doors and Windows for P3190A, $111.00 extra; for P3190B, $78.00 extra.*
*Screen Doors and Windows, galvanized wire, for P3190A, $69.00 extra; for P3190B, $49.00 extra.*
*Side Pergola, $80.00 extra.*
*Rear Pergola, $136.00 extra.*
*Oak Doors, Trim and Floors in living room, dining room and sun room; also Oak Stairs. Maple Floors in kitchen and bathroom for P3190A, $196.00 extra.*
*Oak Doors, Trim and Floors in living room and dining room; also Oak Stairs. Maple Floors in kitchen and bathroom for P3190B, $166.00 extra.*

### What Our Prices Include

At the prices quoted we will furnish all the material to build this seven-room house consisting of:
Lumber; Lath;
**Shingles,** Best Grade Clear Red Cedar Shingles;
**Siding,** Clear Cypress or Clear Red Cedar, Bevel;
**Framing Lumber,** No. 1 Quality Douglas Fir or Pacific Coast Hemlock;
**Flooring,** Clear Grade Douglas Fir or Pacific Coast Hemlock;
**Porch Ceiling,** Clear Grade Douglas Fir or Pacific Coast Hemlock;
**Finishing Lumber;**
**High Grade Millwork** (see pages 110 and 111);
**Interior Doors,** Two Vertical Panel Design of Douglas Fir;
**Trim,** Beautiful Grain Douglas Fir or Yellow Pine;
**Kitchen Cupboards; Medicine Case;**
**Built-In Ironing Board;**
**Windows** of California Clear White Pine;
**40-Lb. Building Paper; Sash Weights;**
**Eaves Trough and Down Spouts;**
**Stratford Design Hardware** (see page 132);
**Paint** for Three Coats Outside;
**Shellac and Varnish** for Interior Doors and Trim.

Complete Plans and Specifications.

Built on a concrete and brick foundation and excavated under entire house.

We guarantee enough material to build this house. Price does not include cement, brick or plaster.

See Description of "Honor Bilt" Houses on Pages 12 and 13.

For prices of Plumbing, Heating, Wiring, Electric Fixtures and Shades see pages 130 and 131.

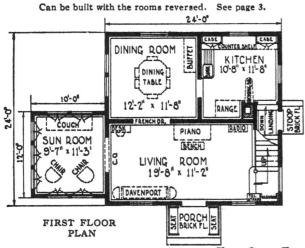

FIRST FLOOR PLAN

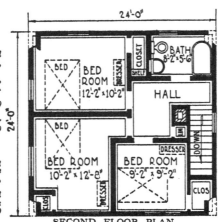

SECOND FLOOR PLAN

*For Our Easy Payment Plan See Page 144*

# THE PURITAN INTERIORS

THE KITCHEN

THE LIVING ROOM

THE
DINING ROOM

THE BEDROOM

THE BATHROOM

All of the furnishings shown above were taken from our Big General Catalog, "The Thrift Book of a Nation."

Gentlemen:-
    I have just recently completed building one of your "Honor Bilt" Modern Homes, and want to tell you how well I am satisfied.  I saved over $2,000.00 in building this house, and when it was completed I was able to get a mortgage for more than the construction cost.  It certainly is a substantial house, and no one will make a mistake in buying or building an "Honor Bilt" Modern Home.
    You might also like to know that it is furnished with Sears-Roebuck rugs, furniture and curtains, also wall paper and fixtures, and in buying my furniture from you I saved over half.  You may refer anyone to me as I know they will be pleased as well as satisfied in dealing with your company.
               (Signed) A. W. Fischer, Eastwood, Ohio

T HE STARLIGHT bungalow is one of our most popular designs. It is dignified and substantial in every detail. Architects and builders say the Starlight has as good an arrangement, considering its size, as it is possible to have. It has the proper number of rooms for the average family. The careful planning, together with our direct-from-factory prices, gives the utmost for the money spent. More than seven hundred have been built.

The main exterior features of this home are the hip roof which extends over the entire house and porch, the dormer, and the porch with its large columns and porch rail. Here in the Starlight you can enjoy comfort in every room, and on a warm day the shady porch bids you welcome for rest, as there is room for hammock, swing and other porch furniture. The porch is 24 feet by 6 feet, and can be glazed or screened in, making it practicable the twelve months in the year.

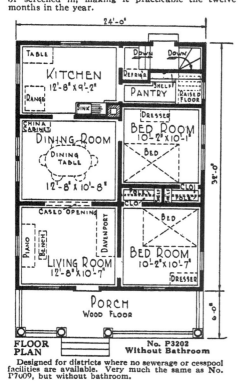

**FLOOR PLAN**      **No. P3202**
**Without Bathroom**

Designed for districts where no sewerage or cesspool facilities are available. Very much the same as No. P7009, but without bathroom.

---

**Honor Bilt**

## The Starlight

**$1,424.00**   No. P3202 (Without Bathroom) "Already Cut" and Fitted.

**$1,542.00**   No. P7009 (With Bathroom) "Already Cut" and Fitted.

**The Living Room.** A three-light door opens into the living room. Size, 12 feet 8 inches by 10 feet 7 inches. There is enough wall space for piano, davenport and other furniture. Two windows and glazed door provide plenty of light and fresh air.

**The Dining Room.** From the living room a cased opening leads into the dining room. This arrangement makes it possible to use the two rooms as one, if occasion requires it. Size of dining room, 12 feet 8 inches by 10 feet 8 inches. A double window provides an abundance of light and ventilation.

**The Kitchen.** A swinging door from dining room leads to the kitchen. Size, 12 feet 8 inches by 9 feet 2 inches. The location for sink, range, table and chairs, has been planned with a view toward helping the housewife reduce needless steps. The kitchen has a shelved pantry in No. P3202 lighted by a window. Kitchen of No. P7009 has a closet with shelf and a pantry. Two windows keep the kitchen bright and well aired. A door leads to rear entry, which has space for ice box, steps to grade and basement. This arrangement permits entrance to basement without going outside.

**The Bedrooms.** There are two bedrooms. The front bedroom opens from the living room. It has a clothes closet and two windows. From the dining room through a cased opening into a small hall, the rear bedroom and bathroom are reached. The rear bedroom, too, has a clothes closet and two windows.

The bathroom has a medicine case, and is lighted by a window.

**Basement.** Excavated basement under entire house. Room for furnace, laundry and storage.

**Height of Ceilings.** Main floor, 9 feet from floor to ceiling. Basement, 7 feet from concrete floor to joists.

### What Our Prices Include

At the prices quoted we will furnish all material to build this five-room bungalow consisting of:
Lumber; Lath;
Roofing, Best Grade Clear Red Cedar Shingles;
Siding, Clear Grade Cypress or Clear Red Cedar, Bevel;
Framing Lumber, No. 1 Quality Douglas Fir or Pacific Coast Hemlock;
Flooring, Clear Grade Douglas Fir or Pacific Coast Hemlock;
Porch Flooring, Clear Grade Edge Grain Fir;
Porch Ceiling, Clear Grade Douglas Fir or Pacific Coast Hemlock;
Finishing Lumber;
High Grade Millwork (see pages 110 and 111);
Interior Doors, Five Cross Panel Design of Douglas Fir;
Trim, Beautiful Grain Douglas Fir or Yellow Pine;
Windows of California Clear White Pine;
Medicine Case for No. P7009;
Eaves Trough and Down Spout;
40-Lb. Building Paper; Sash Weights;
Stratford Design Hardware (see page 132);
Paint for Three Coats Outside Trim and Siding;
Shellac and Varnish for Interior Doors and Trim.
Complete Plans and Specifications.

---

**Can be built on a lot 30 feet wide**

This house can be built with the rooms reversed. See Page 3

**FLOOR PLAN**      **No. P7009**
**With Bathroom**

We guarantee enough material to build this house. Prices do not include cement, brick or plaster. See description of "Honor Bilt" Houses on pages 12 and 13.

For prices of Plumbing, Heating, Wiring, Electric Fixtures and Shades see pages 130 and 131.

### OPTIONS

*Sheet Plaster and Plaster Finish, to take the place of wood lath, $137.00 extra for No. P7009 and $130.00 extra for No. P3202. See page 109.*

*Oriental Asphalt Shingles, instead of wood shingles, $35.00 extra for No. P7009 and $33.00 for No. P3202.*

*Oak Doors, Trim and Floors in living and dining room. Maple Floors in kitchen and bathroom, $109.00 extra for No. P7009 and $110.00 for No. P3202.*

*Storm Doors and Windows, $53.00 extra for No. P7009 and $49.00 for No. P3202.*

*Screen Doors and Windows, galvanized wire, $35.00 extra for No. P7009 and $32.00 extra for No. P3202.*

---

*For Our Easy Payment Plan See Page 144*

---

THE DEL REY bungalow was first built in Pasadena, California, where it was admired by travelers from all parts of the world. Among its many unique features is the wide overhanging roof that shelters the windows and sidewalls from rain and sun and affords special protection to the three long French windows at the front. All windows on front elevation are the long French windows, reflecting a touch of Italian and Spanish architecture. A good size front porch, 20 feet by 8 feet, is partly covered by the extending main roof. Over the doorway there is a separate roof supported by ornamental brackets, carrying out the quaint and beautiful style of architecture.

**Honor Bilt**

## The Del Rey

### No. P13065 "Already Cut" and Fitted
### $2,557.00

**The Living Room** is of large size and perfect proportions, being 21 feet 6 inches by 15 feet 4 inches. An attractive mantel and fireplace is on the right hand side as you enter. On each side of the fireplace is a built-in bookcase, with casement sash above. This large living room not only provides ample space for piano, davenport, phonograph and other furniture, but permits a variety of arrangements. The large French windows provide an abundance of sunshine and fresh air. A coat closet is provided in the corner of living room.

**The Dining Room.** French doors lead from the living room into the dining room. Size, 13 feet 8 inches by 13 feet 6 inches. This room has plenty of space for dining room furniture, with ample room to serve. Two windows give plenty of light and ventilation.

**The Screened Porch** is entered from the dining room or from the rear bedroom. It is available as a summer breakfast room, playroom or sleeping porch.

**The Kitchen.** From the dining room a swinging door opens into the kitchen. Careful study by the architect has planned all work near the windows, and reduced the housewife's many steps to the minimum. Cupboards of the latest design, ample for every need, simplify the daily tasks. There is room for an extra table and chairs. A door opens into the rear entry, which provides for an ice box. There is a door to grade and steps to basement.

**The Bedrooms.** From the living room a hall connects with the two bedrooms and bath. The front bedroom has three French windows on the front and two French windows on the side. Size of room, 13 feet 4 inches by 10 feet 5 inches. It has a three-compartment wardrobe with plate glass mirrors, as illustrated on pages 110 and 111 (No. P9266). The rear bedroom is 11 feet 7 inches by 11 feet 11 inches. It has a clothes closet. Two windows furnish light and air. A door connects with screened porch.

The bathroom is located between the two bedrooms with door facing the hallway. Bathroom plumbing is arranged on one wall. A medicine cabinet and towel case with five adjustable shelves are other features.

**Basement.** Room for furnace, laundry and storage.

**Height of Ceiling.** Main floor, 9 feet from floor to ceiling. Basement, 7 feet from floor to joists.

### What Our Price Includes

At the price quoted we will furnish all the material to build this five-room bungalow, consisting of:
**Lumber; Lath;**
**Roofing,** Oriental Asphalt Shingles, 17-Year Guarantee;
**Siding,** Clear Cypress or Clear Red Cedar, Bevel;
**Framing Lumber,** No. 1 Quality Douglas Fir or Pacific Coast Hemlock;
**Flooring,** Clear Grade Douglas Fir or Pacific Coast Hemlock;
**Porch Ceiling,** Clear Grade Douglas Fir or Pacific Coast Hemlock;
**Finishing Lumber;**
**High Grade Millwork** (see pages 110 and 111);
**Interior Doors,** Two Vertical Panel Design of Douglas Fir;
**Trim,** Beautiful Grain Douglas Fir or Yellow Pine;
**Windows** of California Clear White Pine;
**Screens** for Porch;
**Medicine Case; Kitchen Cabinet;**
**Wardrobes; Mantel;**
**Eaves Trough and Down Spouts;**
**40-Lb. Building Paper; Sash Weights;**
**Chicago Design Hardware** (see page 132);
**Paint** for Three Coats Outside Trim and Siding;
**Shellac and Varnish** for Interior Doors and Trim;
**Shellac, Paste Filler and Floor Varnish** for Oak and Maple Floors;
Complete Plans and Specifications.
We guarantee enough material to build this house. Price does not include cement, brick or plaster.

See description of "Honor Bilt" Houses on pages 12 and 13.

### OPTIONS

*Sheet Plaster and Plaster Finish to take the place of wood lath, $178.00 extra. See page 109.*
*Oak Doors and Trim in living room and dining room.*
*Maple Floors in kitchen and bathroom, $125.00 extra.*
*Storm Doors and Windows, $92.00 extra.*
*Screen Doors and Windows, galvanized wire, $42.00 extra.*

For prices of Plumbing, Heating, Wiring, Electric Fixtures and Shades see pages 130 and 131

**FLOOR PLAN**

40'-0"
36'-0"

SCREENED PORCH 12'-0" x 8'-0"
DINING ROOM
DINING TABLE
CASE
CUPBOARD
KITCHEN 9'-2" x 13'-6"
SINK
SIDEBOARD
RANGE
DOWN
BED ROOM 11'-7" x 11'-11"
DRESSER
13'-8" x 13'-6"
Ice Box
FRENCH DR.
DAVENPORT
CHAIR
CLOS. SHELF
BATH 9'-9" x 5'-11"
HALL
C.O.
PIANO
LIVING ROOM 21'-6" x 15'-4"
TABLE
MANTEL
BOOKCASE
BED
RADIO
CHAIR
BOOKCASE
BED ROOM 13'-4" x 10'-5"
DRESSER
CLOSET
PORCH 20'-0" x 8'-0" CEMENT FLOOR

**Can be built on a lot 50 feet wide**

**Brick Mantel and Bookcases in Living Room**

*For Our Easy Payment Plan See Page 144*

TO THE folks who like a touch of individuality with good taste the Crescent bungalow makes a special appeal. The front door, side lights, and windows have been admirably selected. Seldom, indeed, do you find a more inviting front porch, hood supported by graceful columns, and entrance than we provide for this house. Your choice of two floor plans as shown.

**The Living Room** measures 20 feet 5 inches by 12 feet 5 inches in No. P13084A, and 17 feet 6 inches by 12 feet 2 inches in No. P13086A. Plenty of space for a piano and furniture. The open stairway presents a beautiful effect, and there is a door at the top to prevent drafts from the attic.

**The Dining Room.** The large living room and the dining room are connected by means of a wide cased opening. Floor area of dining room, 12 feet 10 inches by 12 feet 5 inches in No. P13084A, and 12 feet by 10 feet 8 inches in No. P13086A.

**The Kitchen.** Very handy is the swinging door that connects the dining room and the kitchen. The kitchen is equipped with a built-in cupboard, has space for the sink, range, table and chair. Door leads to rear porch, stairs to grade, and basement in No. P13084A, and to side entry and basement in No. P13086A.

**Can be built on a lot 40 feet wide**

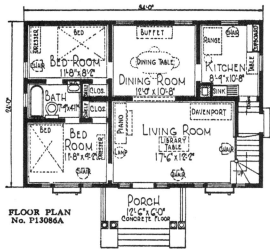

**FLOOR PLAN
No. P13086A**

The Crescent Home is shown in colors on the front cover

**Honor Bilt**

## The Crescent
**$1,900.00** No. P13086A "Already Cut" and Fitted
**2,410.00** No. P13084A "Already Cut" and Fitted

**The Bedrooms.** Either floor plan has two bedrooms with clothes closets, and a bathroom convenient to either bedroom. All bedrooms are well lighted and aired.

**The Basement.** Room for furnace, laundry and storage.

**Height of Ceilings.** Main floor, 9 feet from floor to ceiling. Basement, 7 feet from floor to joists.

### What Our Price Includes

At the price quoted we will furnish all the material to build this five-room bungalow, consisting of:

Lumber; Lath;
Roofing, Oriental Slate Surfaced Shingles, 17-Year Guarantee;
Siding, Clear Cypress or Clear Red Cedar, Bevel; Clear Red Cedar Shingles for Gables;
Framing Lumber, No. 1 Quality Douglas Fir or Pacific Coast Hemlock;
Flooring, Clear Maple for kitchen and bath, Clear Oak for balance of rooms;
Porch Ceiling, Clear Douglas Fir or Pacific Coast Hemlock;
Finishing Lumber;
High Grade Millwork (see pages 110 and 111);
Interior Doors, Inverted Two-Panel Design of Douglas Fir;
Trim, Beautiful Grain Douglas Fir or Yellow Pine;
Windows, California Clear White Pine;
Medicine Case;
Kitchen Cupboard;
Eaves Trough; Down Spout;
40-Lb. Building Paper; Sash Weights;
Stratford Design Hardware (see page 132);
Paint for Three Coats Outside Trim and Siding;
Stain for Two Brush Coats for Shingles on Gable Walls;
Shellac and Varnish for Interior Trim and Doors;
Shellac, Paste Filler and Floor Varnish for Oak and Maple Floors.
Complete Plans and Specifications.
We guarantee enough material to build this house. Price does not include cement, brick or plaster. See description of "Honor Bilt" Houses on pages 12 and 13.

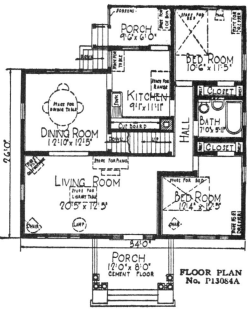

**FLOOR PLAN
No. P13084A**

### OPTIONS

Furnished with two rooms in attic, with single floor, $225.00 extra, for No. P13086A and $182.00 extra for No. P13084A.

Sheet Plaster and Plaster Finish, to take the place of wood lath, for No. P13086A, $161.00 extra; with attic, $214.00 extra; for No. P13084A, $166.00 extra; with attic, $224.00 extra. See page 109.

Storm Doors and Windows, for No. P13086A, $57.00; with attic, $68.00 extra; for No. P13084A, $85.00; with attic, $95.00 extra.

Screen doors and Windows, galvanized wire, for No. P13086A, $34.00; with attic, $40.00 extra; for No. P13084A, $38.00; with attic, $43.00 extra.

Oak Doors and Trim, for living room and dining room, and Oak Stairs for No. P13086A, $123.00 extra. Oak Doors and Trim, for living room and dining room for No. P13084A, $158.00 extra.

For prices of Plumbing, Heating, Wiring, Electric Fixtures and Shades see pages 130 and 131.

*For Our Easy Payment Plan See Page 144*

# The CRESCENT INTERIORS

*ABOVE*—The Pleasant Dining Room.

*BELOW*—The Modern Bathroom.

*These views show one of many ways to furnish The Crescent Home.*

*ABOVE*—The Living Room looking towards the Dining Room and stairs to second floor.

*CENTER*—The Stairway to the second floor.

*BELOW*—The front Bedroom has plenty of light and cross current of air.

*ABOVE*—The Kitchen is equipped with built-in Cupboards.

*BELOW*—Just a glimpse of a Bedroom in the attic. See options.

## The Fairy

### No. P3217 "Already Cut" and Fitted
## $986.00

THE FAIRY is a comfortable bungalow home, with shingle siding. It has many features that will appeal to the housewife. There is quality in every foot of material. Yet, we will furnish all the material to build this bungalow at the very low price quoted. The splendid value is made possible by our modern and successful system of "Honor Bilt" homes.

**The Living and Dining Room.** The porch affords a pleasant place for warm summer evenings. The half glazed front door enables the housewife to see the caller before opening the door.

Having entered you find the living room combines with the dining room. Hence the most economical use of floor space, without sacrifice of appearance. This combination living and dining room is about 12 feet wide by 15½ feet long. There is room for a piano between two windows on the outside wall. Your dining table and chairs may be set at the end of room nearest the kitchen and hidden by a screen if so desired. There is ample space for other furniture as shown on plan.

**The Kitchen.** A swinging door leads from the dining and living room into the kitchen. Here is a well planned kitchen, room to place cabinet, table, sink and range to make work easy, which reduces the housewife's work and saves many steps each day. Two windows give plenty of light and air.

A door connects the kitchen with the rear entry, which has space for ice box, that can be iced without tracking dirt into the kitchen. From the rear entry, stairs lead to the outside and down to basement.

**The Bedrooms.** The front bedroom opens from the living room. It has a fair size clothes closet with shelf.

A small hall is open from the living and dining room and gives privacy to rear bedroom and bath. Bathroom is just the right size and is conveniently located.

**Basement.** Excavated under entire house. Cement floor.

**Height of Ceilings.** Basement is 7 feet from floor to joists. Main floor rooms are 8 feet 6 inches from floor to ceiling.

### What Our Price Includes

At the price quoted we will furnish all the material to build this four-room bungalow, consisting of:

**Lumber,** Red Cedar Shingles for Siding; **Lath;**
**Roofing,** Best Grade Clear Red Cedar Shingles;
**Framing Lumber,** No. 1 Douglas Fir or Pacific Coast Hemlock;
**Flooring,** Interior Floors, Clear Douglas Fir or Pacific Coast Hemlock;
**Porch Floor,** Clear Edge Grain Fir;
**Porch Pergola;**
**Finishing Lumber;**
**High Grade Millwork** (see pages 110 and 111);
**Interior Doors,** Two Cross Panel Designs of Douglas Fir;
**Trim,** Beautiful Grain Douglas Fir or Yellow Pine;
**Windows** of California Clear White Pine;
**Medicine Cabinet; Building Paper; Sash Weights;**
**Eaves Trough and Down Spouts;**
**Stratford Design Hardware** (see page 132);
**Paint** for Three Coats Outside Trim;
**Stain** for Two Coats, for Shingles on Walls;
**Shellac and Varnish** for Inside Doors and Trim.
Complete Plans and Specifications.
Built on concrete foundation, cement blocks above grade.
We guarantee enough material to build this house. Price does not include cement, brick, concrete blocks or plaster.
See description of "Honor Bilt" Houses on pages 12 and 13.

**Can Be Built on a Lot 28 Feet Wide**

This house can be built with the rooms reversed. See page 3.

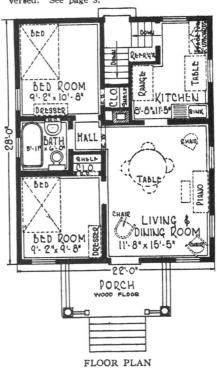

FLOOR PLAN

#### OPTIONS

*Sheet Plaster and Plaster Finish, to take the place of wood lath, $105.00 extra. See page 109.*

*Storm Doors and Windows, $36.00 extra.*

*Screen Doors and Windows, galvanized wire, $25.00 extra.*

*Oriental Asphalt Slate Surfaced Shingles, instead of wood shingles, $27.00 extra.*

For prices of Plumbing, Heating, Wiring, Electric Fixtures and Shades see pages 130 and 131.

*For Our Easy Payment Plan See Page 144*

**Honor Bilt**

## The Oak Park

**$3,265.00** No. P3237B "Already Cut" and Fitted
**$3,012.00** No. P3237A "Already Cut" and Fitted

Can be built on a lot 45 feet wide

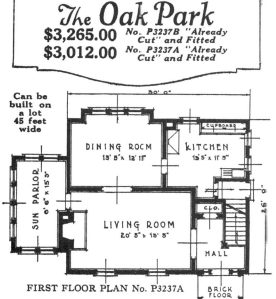

FIRST FLOOR PLAN No. P3237A

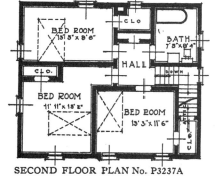

SECOND FLOOR PLAN No. P3237A

***For Our Easy Payment Plan***
***See Page 144***

THE OAK PARK two-story home was designed to reflect the modern trend in Dutch colonial architecture. Here are the self-same beautiful lines that have won the admiration of all generations since Dutch colonial architecture was introduced. Here, too, you can choose from two of the newest and most approved modern interior arrangements.

**The Exterior.** See the beautiful colonial entrance with its sidelights and fan over the door, hood and columns, and the brick porch! Then, again, the colonial windows and green shutters are offset by pure white siding. Its red brick fireplace chimney overlooks the wide overhanging roof.

**INTERIOR No. P3237B →**
**FIRST FLOOR**

**The Reception Hall.** Immediately inside the beautiful entrance is the reception hall where guests are welcomed. A colonial stairway ascends to the second floor. At the rear a door opens to passageway to the kitchen, side entry and stairs to basement. At the left, French doors open into the very pride of the home, the spacious living room.

**The Living Room** floor dimensions are 21 feet 5 inches by 13 feet 5 inches. The colonial fireplace mantel faces the guest as he enters. Good wall space accommodates a piano, and floor proportions provide for satisfactory arrangement of furniture. Two front windows admit light and air.

**The Sun Room.** A single French door leads from the living room to the sun room. Eight windows flood this room with sunshine. Size, 8 feet 6 inches by 15 feet 2 inches.

**The Dining Room.** French doors connect the living room with the dining room. Here is exceptional floor space for a home of this size: 14 feet 5 inches by 12 feet 11 inches. Three windows on the rear wall assure a cheerful atmosphere.

**The Dining Alcove.** Here is a built-in breakfast set and a built-in china cabinet. One window admits light and air.

**The Kitchen.** A wide cased opening unites the dining alcove with the kitchen. Size of the kitchen, 10 feet 5 inches by 11 feet 5 inches. There is a convenient space for the range, the sink, the refrigerator and the handy built-in cupboard. Two windows insure good lighting and airing.

**SECOND FLOOR**

**The Bedrooms.** A hall connects with each of the four bedrooms, bathroom and linen closet.

**The Bathroom** has a built-in medicine cabinet, and a window.

**The Basement.** Room for furnace, laundry and storage.

**Height of Ceilings.** First floor, 8 feet 6 inches from floor to ceiling. Second floor, 8 feet 6 inches from floor to ceiling. Basement, 7 feet from floor to joists.

**← INTERIOR No. P3237A**

The general arrangement interior No. P3237A is the same as interior No. P3237B except that it has no alcove, only three bedrooms. It has a guest closet in the reception hall instead of the passageway to kitchen. For additional details see floor plan on this page.

### What Our Price Includes

At the price quoted we will furnish all material to build this eight-room house, consisting of:

Lumber; Lath;
Roofing, Best Grade Clear Red Cedar Shingles;
Siding, Clear Cypress or Clear Red Cedar, Bevel;
Framing Lumber, No. 1 Quality Douglas Fir or Pacific Coast Hemlock;
Flooring, Clear Grade Douglas Fir or Pacific Coast Hemlock;
Finishing Lumber;
High Grade Millwork (see pages 110 and 111);
Interior Doors, Two-Panel Design Douglas Fir;
Trim, Beautiful Grain Douglas Fir or Yellow Pine;

Windows of California Clear White Pine;
Kitchen Cabinets;
Medicine Case;
China Cabinet for P3237B;
Breakfast Alcove, Table and Seats for P3237B;
Mantel; Colonial Shutters;
Eaves Trough and Down Spout;
40-Lb. Building Paper; Sash Weights;
Chicago Design Hardware (see page 132);
Paint for Three Coats Outside Trim and Siding;
Varnish and Shellac for Interior Doors and Trim.

Can be built on a lot 47 feet wide.

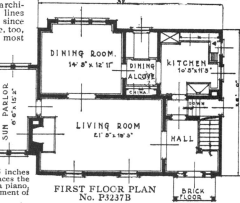

FIRST FLOOR PLAN No. P3237B

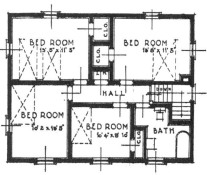

SECOND FLOOR PLAN No. P3237B

**OPTIONS**

*Sheet Plaster and Plaster Finish,* to take the place of wood lath, $255.00 extra for No. P3237A; $274.00 extra for No. P3237B. See page 109.

*Storm Doors and Windows,* $128.00 extra.

*Screen Doors and Windows,* galvanized wire, $81.00 extra.

*Oriental Slate Surfaced Shingles,* in place of wood shingles, for roof, $40.00 extra for No. P3237A; $43.00 extra for No. P3237B.

We guarantee enough material to build this house. Price does not include cement, brick or plaster. See description of "Honor Bilt" Houses on pages 12 and 13.

For prices of Plumbing, Heating, Wiring, Electric Fixtures and Shades see pages 130 and 131.

THE REMBRANDT is an unusually well arranged Dutch Colonial house. It has many special features not generally found in houses of this price. It has a charming entrance that gives an atmosphere of welcome. It has Colonial windows with divided lights above and one light below. French windows in the sun room. Add to this the white siding and contrasting red or green roof with the red brick chimney and you have a home that is sure to charm the most critical.

The interior is cleverly planned. While it has the latest conveniences, the price is unusually low. Why? Because of careful planning and no wasted space. If a house of this size meets with your requirements, you will make no mistake in selecting the Rembrandt.

### FIRST FLOOR

**The Reception Hall.** Entry into the reception hall reveals the splendid character of this Dutch Colonial home. An open stairway leads to the second floor. To the left of stairway is a coat closet with pole for coats and a hat shelf, accommodating a large number of guests' clothes. A grandfather clock can be set along the left wall. On each side of this reception hall are wide cased openings giving an excellent view of the large living room and the dining room on the opposite side.

**The Living Room.** To the right of hall you enter the living room, which is of the proper proportions to accommodate its furnishings to the best advantage. It is unusually large for a house of the size of the Rembrandt. See the floor plan. Almost directly opposite the wide cased opening is a brick mantel which is the central feature. All furniture may be grouped to your liking, as the space is ample. Windows on three sides afford plenty of light and air, making a cheery room.

**The Sun Room.** A French door opens into the sun room from the living room if P3215A is ordered. Seven pairs of French windows flood this room with sunshine. Tastily furnished and draped, it can be made the chosen spot in winter or summer.

### Honor Bilt

## The Rembrandt
### "Already Cut" and Fitted
**P3215A Price with Sun Room, $2,770.00**
**P3215B Price without Sun Room, $2,466.00**

**The Dining Room.** To the left of the hall is the dining room. It is of good size. Windows on two sides give the desired light and ventilation.

**The Kitchen.** From the dining room you pass a swinging door into the model kitchen. Here the sink, the kitchen cabinets, the pantry closet and space for stove and table have been arranged with a thought to save the housewife (or servant) many steps each day. A high sash over the range gives plenty of light over the range and sink. The rear entry provides space for an ice box, which can be iced without iceman tracking the kitchen floor. To the right are stairs leading to the basement and to the left is the grade door to yard.

### SECOND FLOOR

**The Bedrooms.** The stairway hall has a good size window at the landing. A short hall connects with all bedrooms and bath. On one side is the main bedroom of large size which accommodates twin beds, dresser, chiffonier and other furniture. Two clothes closets are at one end of room. Between the closets and underneath a window is a built-in seat, with chest underneath. There are two other bedrooms, each with clothes closets. Ample ventilation and light are assured by the correct number of windows in each room. Close to the bathroom and handy to the bedrooms is a linen closet in the second floor hall.

**The Basement.** Room for furnace, laundry and storage.

**Height of Ceilings.** Basement, 7 feet high from floor to joists, with cement floor. First floor, 9 feet from floor to ceiling. Second floor, 8 feet 7 inches from floor to ceiling.

### What Our Prices Include

At the prices quoted we will furnish all the material to build this six-room colonial house, consisting of:
Lumber; Lath;
Roofing, Oriental Slate Surfaced Shingles, Guaranteed for 17 Years;
Siding, Clear Cypress or Clear Red Cedar, Bevel;
Framing Lumber, No. 1 Quality Douglas Fir or Pacific Coast Hemlock;
Flooring, Clear Grade Douglas Fir or Pacific Coast Hemlock;
Porch Ceiling, Clear Grade Douglas Fir or Pacific Coast Hemlock;
Finishing Lumber; High Grade Millwork (see pages 110 and 111);
Interior Doors, Two Cross Panel Design of Douglas Fir;
Trim, Beautiful Grain Douglas Fir or Yellow Pine, Birch Stair Treads and Rail;
Mantel;
Kitchen Cabinets; Medicine Case;
Windows of California Clear White Pine;
40-Lb. Building Paper; Sash Weights;
Eaves Trough and Down Spout;
Stratford Design Hardware (see page 132);
Paint for Three Coats Outside Trim and Siding;
White Enamel Finish for Trim in Living Room and Hall; Also for Stair Risers and Balusters to Landing;
Mahogany Finish for Doors in Hall, Stair Treads and Rail;
Shellac and Varnish for All Other Interior Trim and Doors.
Complete Plans and Specifications.
Built on a concrete and brick foundation and excavated under entire house.
We guarantee enough material to build this house.
Prices do not include cement, brick or plaster.
See description of "Honor Bilt" Houses on pages 12 and 13.

### OPTIONS

*Sheet Plaster and Plaster Finish, to take the place of wood lath and plaster, with sun room, $242.00; without sun room, $230.00 extra. See page 109.*
*Storm Doors and Windows, with sun room, $114.00 extra; without sun room, $81.00.*
*Screen Doors and Windows, galvanized wire, with sun room, $63.00 extra; without sun room, $51.00.*
For prices of Plumbing, Heating, Wiring, Electric Fixtures and Shades see pages 130 and 131.

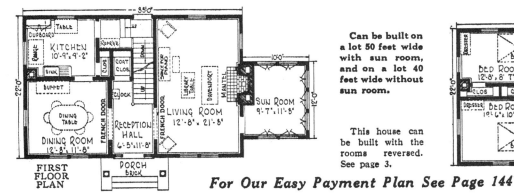

FIRST FLOOR PLAN

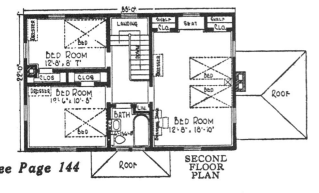

SECOND FLOOR PLAN

Can be built on a lot 50 feet wide with sun room, and on a lot 40 feet wide without sun room.

This house can be built with the rooms reversed. See page 3.

*For Our Easy Payment Plan See Page 144*

# The REMBRANDT ~ INTERIORS

The Living Room, size about 13x22 feet, has an arrangement of charm and dignity

The Reception Hall reveals the fine character of the Rembrandt to the Visitor

The Rembrandt's Dining Room at top of page is — beautiful and simple. It has plenty of sunshine and cheer.

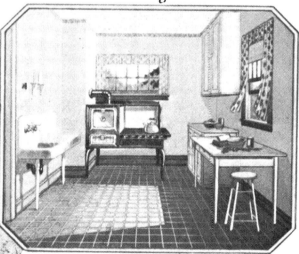

The perfectly equipped Kitchen saves strength and time

A quiet and restful atmosphere of good taste prevails in this spacious Bedroom at top of page

HERE is a masterpiece in a five-room "HONOR BILT" bungalow. The covered porch, size 12 feet by 6 feet, with its massive stucco columns, the stucco gable, the soft toned shingle sides and the wood gable siding, produce a perfect, harmonious effect. The Wellington has been built in many of the choicest locations and is admired wherever it is built. It has found ready sale at a profit of about $2,000.00 to the builder.

By the use of 12-foot studding, the main floor joists are 2 feet 11 inches above the concrete foundation, giving an unusual height to the shingle walls. An artistic touch is added by the flower box beneath the front window and the massive brick chimney on the right elevation.

**The Living Room.** Crossing the porch, you enter the living room through an eight-light glazed door. Size of room is 12 feet 4 inches by 15 feet 7 inches. There is plenty of sunshine and air from three windows. A well designed brick mantel is located on the outside wall and on either side of the fireplace are hinged French windows. The ceiling is ornamented by cornice mold. Here is space for a piano, davenport and other furniture.

**The Dining Room.** The arrangement of the living room and dining room permits an unobstructed view through the wide cased opening, allowing the two rooms to be thrown into one very large room, if so desired. The dining room walls are ornamented with molded panels, used in high class buildings. A large double window provides an abundance of light and air. The size of room is 14 feet 4 inches by 11 feet 3 inches. There is plenty of space to seat the happy family around the dining room table. Space is provided for buffet on the inside wall.

**The Kitchen.** A swinging door leads to the kitchen. This is a well arranged housewife's workroom. Preparing a meal becomes a pleasure because of saving steps due to placing the sink, stove and work table where they belong. The built-in cupboard, where dishes, kitchen utensils and provisions can be stored, is an added feature. Cross ventilation and light are provided by a window on the side, and another in the rear. The door opens into a rear entry, which leads either to the basement or to the outside. Directly opposite the kitchen door, space is provided for refrigerator.

## The Wellington
### No. P3223 "Already Cut" and Fitted
## $1,988.00

**The Bedrooms and Bath.** A small hall, directly off the dining room, connects with two large bedrooms and bathroom—an ideal arrangement. Immediately off the hall is a coat closet. Each bedroom has a closet with hat shelf and wardrobe pole. Each bedroom has two windows, giving light and cross ventilation.

**The Basement.** Basement with cement floor under the entire house. There is ample space for a work bench, laundry, storage and fuel.

**Height of Ceilings.** Basement, 7 feet from floor to joists. Main floor, 8 feet 6 inches from floor to ceiling.

### What Our Price Includes

At the price quoted we will furnish all the material to build this five-room bungalow, consisting of:

Lumber; Lath;
**Roofing,** Best Grade Clear Red Cedar Shingles;
**Siding,** Clear Cypress or Clear Red Cedar, Bevel, Above Belt Course;
**Siding,** Best Grade Thick Cedar Shingles;
**Framing Lumber,** No. 1 Quality Douglas Fir or Pacific Coast Hemlock;
**Flooring,** Clear Maple for Kitchen and Bathroom, Clear Oak for Other Rooms;
**Porch Floor,** Clear Edge Grain Fir;
**Porch Ceiling,** Clear Grade Douglas Fir or Pacific Coast Hemlock;
**Finishing Lumber;**
**High Grade Millwork** (see pages 110 and 111);
**Interior Doors,** Two Vertical Panel Design of Douglas Fir;
**Trim,** Beautiful Grain Douglas Fir or Yellow Pine;
**Kitchen Cupboards; Medicine Case;**
**Brick Mantel;**
**Windows** of California Clear White Pine;
**40-Lb. Building Paper; Sash Weights;**
**Eaves Trough and Down Spout;**
**Chicago Design Hardware** (see page 132);
**Paint,** Three Coats Outside Trim and Bevel Siding;
**Stain,** Two Brush Coats for Shingles on Walls;
**Shellac and Varnish** for Interior Trim and Doors;
**Shellac, Paste Filler and Floor Varnish** for Oak and Maple Floors.
Complete Plans and Specifications.
We guarantee enough material to build this house. Price does not include cement, brick or plaster. See description of "HonorBilt" Houses on pages 12 and 13.

Can be built on a lot 30 feet wide

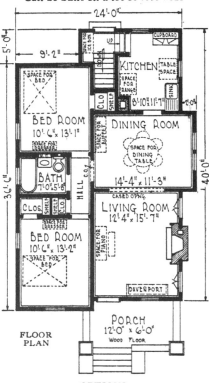

FLOOR PLAN

### OPTIONS

*Sheet Plaster and Plaster Finish to take the place of wood lath, $147.00 extra. See page 109.*

*Oriental Asphalt Shingles, guaranteed 17 years, instead of wood shingles for roof, $48.00 extra.*

*Oak Doors and Trim in living room and dining room, Maple Floors in kitchen and bathroom, $82.00 extra.*

*Storm Doors and Windows, $52.00 extra.*

*Screen Doors and Windows, galvanized wire, $34.00 extra.*

For prices of Plumbing, Heating, Wiring, Electric Fixtures and Shades see pages 130 and 131.

*For Our Easy Payment Plan See Page 144*

THE SHERIDAN is a popular type of bungalow, planned to give the utmost livable space for its size, 28 by 38 feet. The upkeep cost is very small. All the materials are high grade. Porch extends across the entire front of the bungalow and is 26 feet wide by 8 feet deep. It may be screened or glazed and used as a most desirable room.

**FIRST FLOOR**

**The Living Room.** A view of the living room suggests a comfortable home. Plenty of space permits placing of piano, furniture and pictures. Size, 15 feet 2 inches wide by 12 feet 2 inches deep. Room is well lighted and cross ventilated by the double front window and the two casement windows at the side.

**The Dining Room.** Living and dining rooms were planned and may be used as one room or as separate rooms, being connected by a cased opening. Dining room is well lighted by double window and thorough ventilation is assured. There is ample space to seat and serve the family; also space for a buffet.

**Honor Bilt**

## The Sheridan
### No. P3224 "Already Cut" and Fitted
### $2,245.00

**The Kitchen.** The kitchen is entered from dining room by a swinging door. Sink is immediately inside the door, with space for range alongside. Table space is arranged under double window. Windows afford light and ventilation. A door connects with pantry that is equipped with shelves and lighted by window. There is space for refrigerator, iced from door in entry. A door also opens to entry where steps lead to outdoors and to basement.

**The Bedrooms.** Through an open passage you enter a conveniently located hall leading to bedrooms and stairway to second floor. In designing the hall and bedrooms attention was given to the creation of service closets. A coat closet is located in passage from dining room to hall, with a linen closet in the hall. Front bedroom has ample space for furniture and a clothes closet with pole and shelf. Abundant light and air come from two windows. Rear bedroom has one clothes closet with shelf and wardrobe pole and another closet with shelf. Lighted and ventilated by two windows.

**The Bathroom.** Bathroom is conveniently located between the two bedrooms. A medicine case with plate glass mirror is furnished.

**SECOND FLOOR**

An enclosed stairway leads to second floor. The plan calls for three bedrooms and clothes closets for just the small cost of finishing them. See option under floor plan.

**Basement.** Room for furnace, laundry and storage.

**Height of Ceilings.** Main floor, 8 feet 7 inches, floor to ceiling. Second floor, 8 feet 2 inches, floor to ceiling. Basement, 7 feet, concrete floor to ceiling joist.

### What Our Price Includes

At the price quoted we will furnish all of the material to build this five-room bungalow, consisting of:

Lumber; Lath;
**Roofing,** Oriental Slate Surfaced Shingles;
**Siding,** Clear Cypress or Clear Red Cedar, Bevel;
**Framing Lumber,** No. 1 Quality Douglas Fir or Pacific Coast Hemlock;
**Flooring,** Clear Grade Douglas Fir or Pacific Coast Hemlock;
**Porch Flooring,** Clear Edge Grain Fir;
**Porch Ceiling,** Clear Grade Douglas Fir or Pacific Coast Hemlock;
**Finishing Lumber;**
**High Grade Millwork** (see pages 110 and 111);
**Interior Doors,** Two Vertical Panel Design of Douglas Fir;
**Trim,** Beautiful Grain Douglas Fir or Yellow Pine;
**Medicine Case;**
**Windows** of California Clear White Pine;
**40-Lb. Building Paper; Sash Weights;**
**Eaves Trough; Down Spout;**
**Chicago Design Hardware** (see page 132);
**Paint for Three Coats Outside Trim and Siding;**
**Shellac and Varnish** for Interior Trim and Doors.

Built on a concrete foundation and excavated under entire house.

We guarantee enough material to build this house. Price does not include cement, brick or plaster.

Complete plans and specifications.

"Honor Bilt" construction explained on pages 12 and 13.

#### OPTIONS

*Sheet Plaster and Plaster Finish, to take the place of wood lath, $174.00 extra. With attic, $271.00 extra. See page 109.*

*Storm Doors and Windows, $56.00 extra. With attic, $69.00 extra.*

*Screen Doors and Windows, galvanized, $37.00 extra. With attic, $47.00 extra.*

*Oak Doors, Trim and Floors for living and dining room, Maple Floors in kitchen and bathroom, $142.00 extra.*

For Prices of Plumbing, Heating, Wiring, Electric Fixtures and Shades see pages 130 and 131.

**Can be built on a lot 32 feet wide**

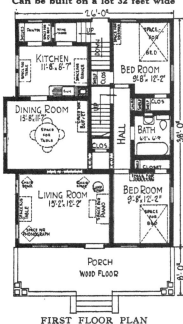

FIRST FLOOR PLAN

Partial view showing living room and dining room of the Sheridan. There are many other ways to furnish this bungalow home.

*For Our Easy Payment Plan See Page 144*   Finished Second Floor Plan, $241.00 Extra

**Can be built on a lot 35 feet wide**

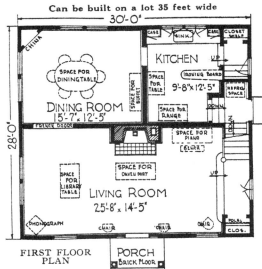

FIRST FLOOR PLAN

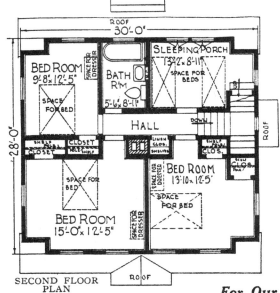

SECOND FLOOR PLAN

THE PRISCILLA will delight lovers of colonial architecture. The entrance with its colonial columns, door and side lights is most inviting, flanked by colonial windows on either side. To the visitor it impresses hospitality that is characteristic of many famous historical homes. The pure white wide siding is in contrast with the sea green or red shingles, green shutters and red brick chimney. Fancy this home in its proper landscaped red tile approach to the brick porch ornamented with flowers and shrubbery. Study the floor plans. Compare the size of rooms with houses you are familiar with of much larger dimensions and you will appreciate the unusual planning of the Priscilla.

### FIRST FLOOR

**The Living Room.** Opening the colonial door you enter a spacious living room extending the entire width of the house. Size, 25 feet 8 inches by 14 feet 5 inches. Directly in front you find a colonial brick mantel. To your left is a double window. To the right an artistic open stairway to the second floor with a coat closet on the landing. In this big room you have ample space for a large piano, phonograph, davenport, library table, overstuffed chairs, reading lamp and other furniture. A French window throws light over the stairway landing. Two front windows look out over the lawn. To the left of the brick mantel is a pair of French doors leading into the dining room.

**The Dining Room.** Size, 15 feet 7 inches by 12 feet 5 inches. Space for a buffet, serving table and tea wagon, as well as dining table and chairs. A china cabinet is built in a corner. See illustration on opposite page. Light and air are provided by two side windows and two rear windows.

**The Kitchen.** A swinging door leads from the dining room to the modern kitchen, which is the particular pride of every Priscilla owner. Our Kitchen De Luxe outfit, disappearing ironing board, the stairway which enables one to reach the second floor from either living room or kitchen, the refrigerator space on kitchen stair landing inside the house but outside of the kitchen, and the modern grade entrance are unusual and most desirable features. Plenty of space for stove and table. Plenty of light and air from two pairs of French casement windows over our Kitchen De Luxe outfit, and one casement window on the stair landing to the second floor.

### SECOND FLOOR

**The Bedrooms.** A hall connects with three bedrooms, sleeping porch, bathroom and linen closet. Each of the two front bedrooms has two front windows and one side window. The rear bedroom has one side window and one rear window. Each bedroom has a clothes closet. The bedroom immediately to the left of stairway has two closets. The sleeping porch has room for two beds. Two pairs of French windows flood this porch with sunshine. Bathroom plumbing is roughed-in on one wall, saving much of the installation cost.

**The Attic.** Stairway leads from second floor to attic. This floor may be used for various purposes.

**Height of Ceilings.** First floor, 9 feet from floor to ceiling. Second floor, 8 feet 6 inches from floor to ceiling. Basement, 7 feet from floor to ceiling.

**Honor Bilt**

# The Priscilla
### No. P3229 "Already Cut" and Fitted
## $3,198.00

### What Our Price Includes

At the price quoted we will furnish all the material to build this six-room and sleeping porch house, consisting of:

**Lumber; Lath;**
**Roofing,** Oriental Asphalt Shingles, 17-Year Guarantee;
**Siding,** Clear Grade Cypress or Red Cedar, Bevel;
**Framing Lumber,** No. 1 Quality Douglas Fir or Pacific Coast Hemlock;
**Flooring,** Clear Oak for Living Room and Dining Room, Clear Maple for Kitchen and Bathroom; Balance of Rooms Clear Grade Douglas Fir or Pacific Coast Hemlock;
**Porch Ceiling,** Clear Grade Douglas Fir or Pacific Coast Hemlock;
**Finishing Lumber;**
**High Grade Millwork** (see pages 110 and 111);
**Interior Doors,** One Panel Design of Douglas Fir;
**Trim,** Beautiful Grain Douglas Fir or Yellow Pine;
**Windows** of California Clear White Pine;
**Medicine Case;** Colonial China Cabinet;
**De Luxe Kitchen Outfit;**
**Built-In Ironing Board;**
**Mantel;**
**Colonial Shutters;**
**Eaves Trough and Down Spouts;**
**40-Pound Building Paper; Sash Weights;**
**Chicago Design Hardware** (see page 132);
**Paint** for Three Coats Outside Trim and Siding;
**Mahogany Stain and Varnish** for All Doors, Stair Treads and Stair Rail; All Other Interior Trim and Woodwork White Enamel.

Complete Plans and Specifications.

We guarantee to furnish enough material to build this house. Price does not include cement, brick or plaster. See description of "Honor Bilt" Houses on pages 12 and 13.

### OPTIONS

*Sheet Plaster and Plaster Finish, to take the place of wood lath, $258.00 extra. See page 109.*
*Cedar Shingles, instead of Oriental Asphalt Shingles, $42.00 less.*
*Storm Doors and Windows, $85.00 extra.*
*Screen Doors and Windows, galvanized wire, $58.00 extra.*

For prices of Plumbing, Heating, Wiring, Electric Fixtures and Shades, see pages 130 and 131.

*For Our Easy Payment Plan See Page 144*

# The PRISCILLA INTERIORS

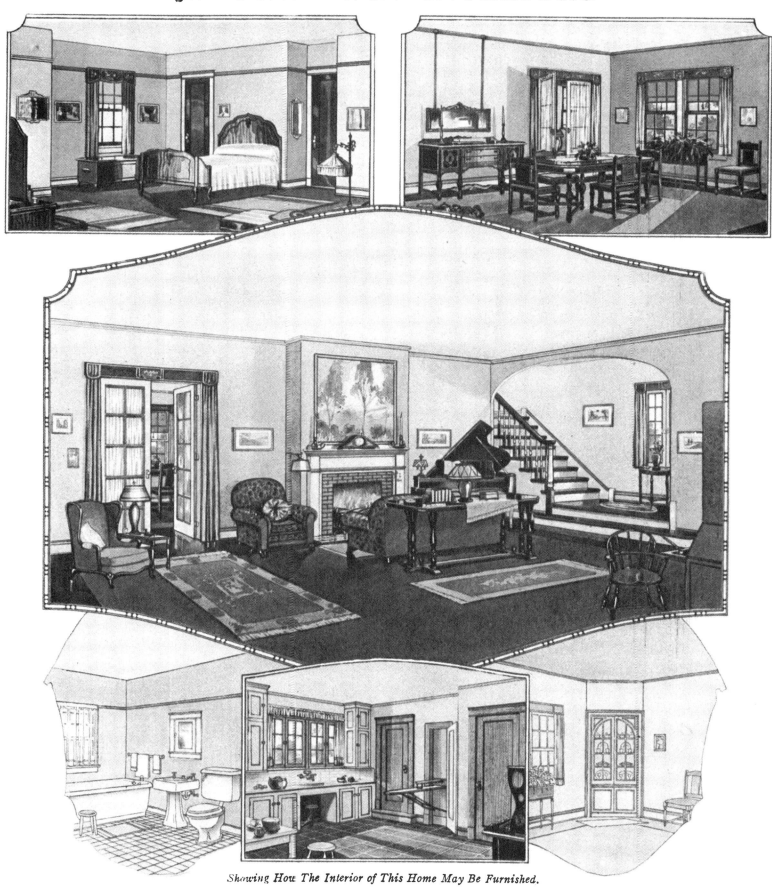

*Showing How The Interior of This Home May Be Furnished.*

*ABOVE*—A Bedroom.

*CENTER*—The spacious Living Room looking towards the Dining Room.

*BELOW*—A glimpse of the Bathroom.

*ABOVE*—The beautiful Dining Room.

*LEFT AT CENTER*—The Kitchen showing our De Luxe Outfit and Built-In Ironing Board.

*BELOW*—A corner of the Dining Room which reveals our Built-In Colonial China Cabinet.

P602   **See Description of the Priscilla Home on Opposite Page**   *Page 35*

**Can be built on a lot 22 feet wide**

FIRST FLOOR PLAN

**Honor Bilt**

## The Norwood
### No. P2095 "Already Cut" and Fitted
### $1,667.00

THE NORWOOD reflects the architecture of a Swiss chalet. Aside from being attractive, it is one of the largest "little houses" designed. The floor plan shows that all rooms are of good size. Due to careful planning there isn't an inch of waste space in the entire house. Imagine this pretty home with its soft tone brown shingles above the belt course, covering the entire second story, with its massive brackets and cornice painted in pure white! The first story is covered with drop siding, that can be painted in a contrasting color, producing a most beautiful effect. The porch, with its overhanging eaves, harmonizes with the gables on main roof. Trellises furnished for both front and right side elevations, add to its charm. The floor area of the first story is 16 feet wide by 32 feet long.

### FIRST FLOOR

**The Living Room.** A ten-light glazed door leads from the porch to the living room, well lighted and ventilated by two front and one side windows. Size, 12 feet 2 inches wide by 11 feet 2 inches deep. There is space for living room furniture, including piano. At left an open stairway leads to the second floor.

**The Dining Room.** Through a cased opening you enter the dining room, 15 feet 3 inches wide by 10 feet deep. There is ample room for furniture, including a buffet. A closet is built in dining room; off this room is the entry to basement. Plenty of sunshine and air are admitted by two windows.

**The Kitchen.** From the dining room a swinging door opens into the kitchen. Space is provided for a range and table, located convenient to the sink. A door opens to pantry. It is provided with shelves and space for refrigerator, with provision for icing it from the outside, doing away with the tracks of the iceman. A door leads to the back porch.

### SECOND FLOOR

**The Bedrooms.** Stairway to second floor connects with a hall which is lighted by a window. The hall gives entry to two bedrooms and bath. Front bedroom is 11 feet 10 inches wide by 11 feet 2 inches deep and is lighted by one double and one side window, giving plenty of light and cross currents of air. The rear bedroom, 15 feet 3 inches wide and 8 feet 8 inches deep, is also lighted and ventilated by two windows, one on the rear and the other on the side. A closet with shelf is provided for each bedroom.

**The Bathroom.** This room is 5 feet 10 inches wide by 8 feet 8 inches long, conveniently located between the two bedrooms. A medicine case with French plate mirror is provided, and liberal space for bathroom fixtures.

**The Basement.** Grade entrance to first floor and basement is from the left side. Plenty of room for furnace, laundry and storage.

**Height of Ceilings.** First floor, 8 feet 4 inches, floor to ceiling. Second floor, 8 feet 2 inches, floor to ceiling. Basement, 7 feet, concrete floor to ceiling joists.

ROOF

SECOND FLOOR PLAN

### OPTIONS

*Sheet Plaster and Plaster Finish, to take the place of wood lath, $161.00 extra. See page 109.*

*Oriental Asphalt Shingles, guaranteed 17 years, instead of wood shingles, $26.00 extra.*

*Storm Doors and Windows, $59.00 extra.*

*Screen Doors and Windows, galvanized wire, $41.00 extra.*

For prices of Plumbing, Heating, Wiring, Electric Fixtures and Shades, see pages 130 and 131.

### What Our Price Includes

At the price quoted, we will furnish all the material to build this five-room two-story house, consisting of:

**Lumber; Lath;**
**Siding,** Clear Grade Cypress or Clear Red Cedar, First Floor;
**Siding,** Best Grade Thick Cedar Shingles, Second Floor;
**Roofing,** Best Grade Clear Red Cedar Shingles;
**Framing Lumber,** No. 1 Quality Douglas Fir or Pacific Coast Hemlock;
**Flooring,** Clear Grade Douglas Fir or Pacific Coast Hemlock for Interior Floor; Clear Edge Grain Fir for Porch;
**Porch Ceiling,** Clear Grade Douglas Fir or Pacific Coast Hemlock;
**Finishing Lumber;**
**High Grade Millwork** (see pages 110 and 111);
**Interior Doors,** Five Cross Panel Design of Douglas Fir;
**Trim,** Beautiful Grain Douglas Fir or Yellow Pine;
**Medicine Case;**
**Windows,** California Clear White Pine;
**40-Lb. Building Paper; Sash Weights;**
**Eaves Trough and Down Spout;**
**Stratford Design Hardware** (see page 132);
**Paint** for Three Coats Outside Trim and Siding;
**Stain** for Shingles on Walls for Two Brush Coats;
**Shellac and Varnish** for Interior Trim and Doors.
Complete Plans and Specifications.
Built on concrete foundation and excavated under entire house.
We guarantee enough material to build this house. Price does not include cement, brick or plaster.
See description of "Honor Bilt" Houses on pages 12 and 13.

*For Our Easy Payment Plan See Page 144*

IMAGINE this cozy home with the siding painted a dark tone color, with rich brown stained shingles for the second story, with window casings painted in pure white!

It is a roomy house at a very low cost, because it is planned on square lines, permitting every square inch of space to be used to the best advantage.

The large porch, size 18 feet by 7 feet, with its brick columns beneath and triple columns above; the wide trusses that support porch roof and the numerous divided light windows, all of the latest design, make this house one of our most popular sellers. The Cornell is the house of sunshine. There is an abundance of light and air in every room. The quality and construction is "HONOR BILT" throughout.

**FIRST FLOOR**

**The Living Room.** From the porch you enter the living room through a handsome glazed door. Towards the right an open stairway rises to the second floor. A hat closet with wardrobe pole and shelf is located on stairway platform. Living room space affords an ideal arrangement for furniture and piano. Five windows supply light and air. Living room is 15 feet 11 inches wide by 12 feet 6 inches deep.

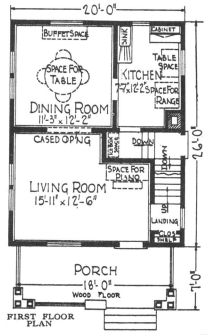

FIRST FLOOR PLAN

**Honor Bilt**

## The Cornell

No. P3226 "Already Cut" and Fitted

### $1,785.00

**The Dining Room.** A cased opening leads to dining room and makes possible the use of living room and dining room as one. Dining room is 11 feet 3 inches deep by 12 feet 2 inches wide. Plenty of space for dining room furniture, with buffet under sash. Sunshine and fresh air are admitted by the large double window at side and the divided light sash at rear.

**The Kitchen.** Kitchen arrangement saves time and effort. Sink is on the left as you enter the kitchen, cupboard near by. Space for a table is located under side window and range space within reach. Kitchen is 7 feet 7 inches wide by 12 feet 2 inches deep. From kitchen a door leads to grade entry and to basement stairway. Refrigerator space is provided in entry.

**SECOND FLOOR**

**The Bedrooms.** A service hall leads to four bedrooms and bath on second floor. The right hand front bedroom is 9 feet 10 inches wide by 9 feet 5 inches deep, and has clothes closet with wardrobe pole and shelf. Adjoining is another bedroom 9 feet wide by 9 feet 4 inches deep. Each rear bedroom has a clothes closet with wardrobe pole and shelf. Each bedroom is exceptionally well lighted and cross ventilated by two windows.

**The Bathroom.** It has a medicine case with plate glass mirror and space for the customary bathroom fixtures. Bathroom is 5 feet 1 inch wide and 5 feet 4 inches deep.

**Basement.** Excavated basement with concrete floor. Room for furnace, laundry and storage.

**Height of Ceilings.** First floor, 8 feet 8 inches, floor to ceiling. Second floor, 8 feet 8 inches, floor to ceiling. Basement, 7 feet, concrete floor to joists.

**OPTIONS**

*Sheet Plaster and Plaster Finish, to take the place of wood lath, $179.00 extra. See page 109.*

*Oriental Asphalt Shingles, guaranteed 17 years, instead of wood shingles, $23.00 extra.*

*Storm Doors and Windows, $72.00 extra.*

*Screen Doors and Windows, galvanized wire, $47.00 extra.*

*Oak Doors, Trim and Flooring in living and dining room, Maple Floors in kitchen and bathroom, $160.00 extra.*

For Prices of Plumbing, Heating, Wiring, Electric Fixtures and Shades, see pages 130 and 131.

**Can be built on a 25-foot lot**

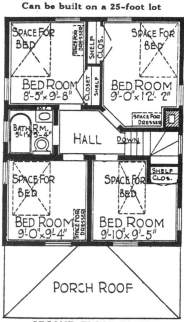

SECOND FLOOR PLAN

**What Our Price Includes**

At the price quoted we will furnish all the material to build this seven-room house consisting of:
Lumber; Lath;
**Roofing,** Best Grade Clear Red Cedar Shingles;
**Siding,** Clear Cypress or Clear Red Cedar, Bevel, Up to Belt Course; Above Belt, Clear Red Cedar Shingles;
**Framing Lumber,** No. 1 Quality Douglas Fir or Pacific Coast Hemlock;
**Flooring,** Clear Grade Douglas Fir or Pacific Coast Hemlock;
**Porch Flooring,** Clear Edge Grain Fir;
**Porch Ceiling,** Clear Douglas Fir or Pacific Coast Hemlock;
**Finishing Lumber;**
**High Grade Millwork** (see pages 110 and 111);
**Interior Doors,** Inverted Two-Panel Design of Douglas Fir;
**Trim,** Beautiful Grain Douglas Fir or Yellow Pine;
**Kitchen Cabinet; Medicine Case;**
**40-Lb. Building Paper; Sash Weights;**
**Eaves Trough; Down Spout;**
**Stratford Design Hardware** (see page 132);
**Paint for Three Coats** Outside Trim and Siding;
**Stain for Shingles** on Walls for Two Brush Coats;
**Shellac and Varnish** for Interior Trim and Doors;
Complete Plans and Specifications.
Built on concrete foundation and excavated under entire house.
We guarantee enough material to build this house. Price does not include cement, brick or plaster.
See description of "Honor Bilt" Houses on pages 12 and 13.

*For Our Easy Payment Plan See Page 144*

**Honor Bilt**

## The Montrose
### No. P3239 "Already Cut" and Fitted
### $2,923.00

THE MONTROSE is justly considered a beautiful home in any community, no matter how exclusive. Assuredly handsome and substantial, it is a rare tribute to the best of English colonial architecture. Judges of good homes proclaim The Montrose as distinctive and charming. It is conveniently appointed, well planned and economical of upkeep.

**The Exterior** is adorned by a long projection around the middle of the house, varied only by the unusually lovely and hooded vestibuled entrance. There is the red brick fireplace chimney, colonial windows with green shutters, except for the sunroom; flower boxes on dining room and front living room windows, wide colonial siding and brick porch.

The Montrose is modern throughout and represents the experience of many years of high grade residence designing. The price is consistently low because of our ready-cut system and elimination of the middleman.

#### FIRST FLOOR
**The Vestibule** has a closet for wraps.
**The Living Room.** You enter the spacious living room from the vestibule. Size, 12 feet 2 inches by 23 feet 3 inches. It extends from the front of house to the rear. Centrally placed in the left wall is a colonial mantel and fireplace. Lighted by a triple window in the rear and one window in the front. A passageway leads to the second floor, and connects with the kitchen.
**The Sunroom.** A French door connects the living room with the sunroom. Floor dimensions of sunroom are: 9 feet 7 inches by 17 feet 3 inches. Twelve windows make this room the brightest in the house.
**The Dining Room.** French doors open into the dining room from the living room. Floor area of dining room is 12 feet 8 inches by 12 feet 3 inches. Being practically square it makes a good setting for a dining set and gives the greatest ease in serving. A double window in the front and a single window on the side insure light and ventilation.

**The Kitchen.** A swinging door in the dining room opens into the modern and efficient kitchen. It is equipped with a built-in kitchen cupboard, opposite which is space for a sink and range; space for table immediately to the right of door leading to dining room door. Size, 9 feet 2 inches by 10 feet 7 inches. A double window admits light and ventilation.
A door on the left leads to the passageway which has stairs to the second floor, thereby making it convenient to reach the second floor or the front door from the kitchen. A door in the rear leads to the enclosed rear entry, which has space for refrigerator, stairs to basement and door to grade.

#### SECOND FLOOR
**The Bedrooms.** The stairway off the passageway in the living room leads to the hall on the second floor. This hall connects with three bedrooms, bathroom, linen closet and door to attic stairs. Each bedroom has a clothes closet and cross ventilation. The bathroom has a built-in medicine case and a window. The linen closet is centrally located, close to bedrooms and bathroom.
**The Basement.** Space for heating plant, laundry and storage.
**Height of Ceilings.** First floor, 9 feet 6 inches from floor to ceiling. Second floor, 8 feet 6 inches from floor to ceiling. Basement, 7 feet from floor to joists. Attic, 9 feet from floor to ridge.

#### What Our Price Includes
At the price quoted we will furnish all the material to build this seven-room two-story home, consisting of:
Lumber; Lath;
**Roofing,** Best Grade Clear Red Cedar Shingles;
**Siding,** Clear Cypress or Clear Red Cedar Bevel;
**Framing Lumber,** No. 1 Quality Douglas Fir or Pacific Coast Hemlock;
**Flooring,** Clear Grade Douglas Fir or Pacific Coast Hemlock;
**Finishing Lumber;**
**High Grade Millwork** (see pages 110 and 111);
**Interior Doors,** Two Panel Design of Douglas Fir;
**Trim,** Beautiful Grain Douglas Fir or Yellow Pine;
**Windows** of California Clear White Pine;
**Medicine Case; Kitchen Cabinets;**
**Mantel; Colonial Shutters;**
**Eaves Trough and Down Spout;**
**40-Lb. Building Paper; Sash Weights;**
**Chicago Design Hardware** (page 132);
**Paint** for Three Coats Outside Trim and Siding;
**Varnish and Shellac** for Interior Doors and Trim.

**Can be built on a lot 42 feet wide**

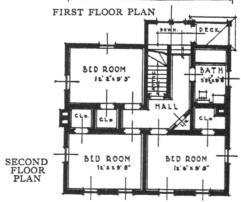

**FIRST FLOOR PLAN**

**SECOND FLOOR PLAN**

We guarantee enough material to build this house. Price does not include cement, brick or plaster.
See description of "Honor Bilt" Houses on pages 12 and 13 of Modern Homes Catalog.

#### OPTIONS
*Sheet Plaster and Plaster Finish, to take the place of wood lath.* $213.00 extra. See page 109.
*Storm Doors and Windows,* $162.00 extra.
*Screen Doors and Windows,* galvanized wire, $109.00 extra.
*Oriental Slate Surfaced Shingles, in place of wood shingles, for roof,* $39.00 extra.

For Prices of Plumbing, Heating, Wiring, Electric Fixtures and Shades see pages 130 and 131.

*For Our Easy Payment Plan See Page 144*

THE LA SALLE, colonial duplex or income bungalow, makes a worth while and high grade investment. It provides: (1) A home for the owner; (2) A steady income from rent of the other apartment. In the long run the owner finds such rental actually pays for the investment. That's why the income bungalow is such a decided success.

The La Salle readily wins approval because of its classical and practical lines. It has conveniences generally found in higher priced bungalows.

**The Exterior** has a colonial front porch, a sloping roof that continues over the porch, colonial windows and shutters. To a passerby The La Salle appears like an exclusive one-family residence.

**Can be built on a lot 30 feet wide**

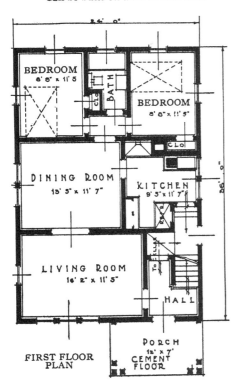

FIRST FLOOR PLAN

### FIRST FLOOR APARTMENT
**The Reception Hall.** From the front porch you enter the reception hall, which is planned to be used for both apartments. It contains the main stairs to the second floor apartment, a door to the passageway, so that either family can reach the basement, or side entry.

**The Living Room** is of good size, measuring 16 feet 2 inches by 11 feet 5 inches, and so readily adapts itself to any furniture arrangement. Light and air are available from a triple window in the front and a high sash on the side.

**The Dining Room.** A pair of French doors divides the living room and the dining room. Size, 13 feet 5 inches by 11 feet 7 inches. Lighted by a double window.

**The Kitchen.** To the right of the dining room a swinging door connects with the kitchen. Floor space measures 9 feet 5 inches by 11 feet 7 inches. Contains place for range, sink and refrigerator, opposite which is a built-in cupboard. A double window provides light and ventilation. A door connects with the side entry, which has stairs to basement, and door to grade.

**The Bedrooms.** Directly off the dining room is a small hall that connects with the two bedrooms and bathroom. Each bedroom has a clothes closet, and two windows with cross ventilation.

**The Bathroom** has a built-in medicine case, and a window.

### SECOND FLOOR APARTMENT
A door at the top of stairs connects with a hall which serves the living room, dining room and kitchen, bedroom and bath.

**The Living Room** measures 16 feet 2 inches by 11 feet 5 inches. It contains a spacious closet for a wall bed, giving a two-bedroom efficiency in a one-bedroom home. Three front windows flood the room with sunshine and air.

**The Dining Room and Kitchen.** This is really one room but the appearance of two rooms is obtained by the use of china cabinets, that divide the dining room and the kitchen as shown on floor plan. Size of dining room is 7 feet 11 inches by 11 feet 11 inches. It has a large closet, and is lighted by a double window. As stated above, two china cabinets form the opening to the kitchen. Size of kitchen, 7 feet 5 inches by 9 feet 8 inches. There is a space allotted for the sink, range, refrigerator, and containing built-in cupboards. A double window over sink provides light and ventilation.

**The Bedroom** measures 11 feet 5 inches by 13 feet 3 inches, and so accommodates twin beds. Has a roomy clothes closet and a double window.

**The Bathroom.** Centrally located off the hall. Has a built-in medicine case and a window.

**The Basement.** Room for a furnace for both apartments or a furnace for each apartment; also laundry and storage.

**Height of Ceilings.** First floor, 8 feet 6 inches from floor to ceiling. Second floor, 8 feet 6 inches from floor to ceiling. Basement, 7 feet from floor to joists.

### What Our Price Includes
At the price quoted, we will furnish all the material to build this nine-room house, consisting of:

Lumber; Lath;

Roofing, Best Grade Clear Red Cedar Shingles;

Siding, Clear Cypress or Clear Red Cedar Bevel;

Framing Lumber, No. 1 Quality Douglas Fir or Pacific Coast Hemlock;

Flooring, Clear Grade Douglas Fir or Pacific Coast Hemlock;

Porch Flooring, Clear Grade Douglas Fir or Pacific Coast Hemlock;

Porch Ceiling, Clear Grade Douglas Fir or Pacific Coast Hemlock;

Finishing Lumber; High Grade Millwork (see pages 110 and 111);

Interior Doors, Two Panel Design of Douglas Fir;

Trim, Beautiful Grain Douglas Fir or Yellow Pine;

Windows of California Clear White Pine;

Medicine Case; Kitchen Cabinets;

Colonial Shutters;

Eaves Trough and Down Spout;

40-Lb. Building Paper; Sash Weights;

Chicago Design Hardware (page 132);

Paint for Three Coats Outside Trim and Siding;

Varnish and Shellac for Interior Doors and Trim.

We guarantee enough material to build this house. Price does not include cement, brick or plaster.

See description of "Honor Bilt" Houses on pages 12 and 13 of Modern Homes Catalog.

*For Our Easy Payment Plan See Page 144.*

## The La Salle
### No. P3243 "Already Cut" and Fitted
## $2,746.00

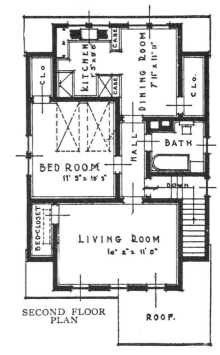

SECOND FLOOR PLAN

### OPTIONS
*Sheet Plaster and Plaster Finish, to take the place of wood lath, $270.00 extra. See page 109.*
*Storm Doors and Windows, $115.00 extra.*
*Screen Doors and Windows, galvanized wire, $75.00 extra.*
*Oriental Slate Surfaced Shingles, in place of wood shingles, $33.00 extra.*
For prices of Plumbing, Heating, Wiring Electric Fixtures and Shades see pages 130 and 131.

# THE ALHAMBRA

THE ALHAMBRA is a modern adaptation of mission architecture. The exterior details call for a stucco siding with a course of red brick above ground, and red or green roof. Windows, doors and decorative molds can be painted in either white, green or any contrasting color. The porch is roofed over the entrance and terraced at the side. While the exterior of the Alhambra is beautiful, its interior is unusually good.

From $2,500 to $4,000 profit has frequently been made by our customers who have built this house, which proves the big value we give in material and the unusual architecture reflected in this design.

**Honor Bilt**

## The Alhambra
### No. P17090A "Already Cut" and Fitted
## $3,134.00

## FIRST FLOOR

**The Living Room and Sun Room.** Entering the Alhambra you readily approve the unique way in which the rooms are arranged. With a foundation of only 28 feet by 31 feet 6 inches, the living room dimensions are 18 feet 9 inches by 14 feet 9 inches. Toward the right a handsome stairway leads to the second floor. The stairway has a wide landing, in the corner of which is a coat closet. In the living room a real fireplace and mantel with a built-in bookcase on each side is the central feature. Ample space for all furniture. Four pairs of French windows provide plenty of light and ventilation.

From the living room a large cased opening leads into the sun room. The sun room measures 8 feet 1 inch by 14 feet 9 inches. It has two double French windows set in a beautiful recessed bay, in which is a built-in seat with hinged seat cover, allowing space for various articles. Two French windows face the open terraced porch.

**The Dining Room.** A pair of French doors connect sun room with the dining room. Space for buffet is provided by a partly recessed nook. Two pairs of French windows in the rear and side wall provide light and ventilation.

### Here You Will Find a Prize Winner in a Model Kitchen

From the dining room a swinging door opens into the kitchen. Along the rear wall are two windows over the sink with a closet on each side to serve as a pantry. The inside walls provide space for gas range, table and chair. In the wall is a built-in ironing board, concealed by a panel door when not in use.

The kitchen has stairway that meets the front stairway from the living room on the upper landing. There they merge into one stairway to second floor. Other steps lead from the kitchen to grade or outside entrance which has space for a refrigerator, stairs to basement and sash door.

**Can be built on a lot 37 feet wide**

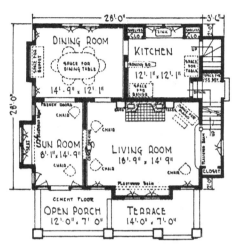

FIRST FLOOR PLAN

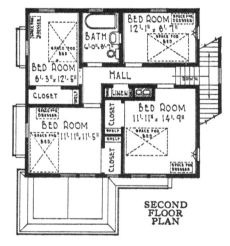

SECOND FLOOR PLAN

## SECOND FLOOR

**The Bedrooms.** A hall leads to four bedrooms, bath and linen closet. Each of the two front bedrooms has a large clothes closet with clothes pole, double doors and shelf above. The rear bedroom at the left has a good size clothes closet with clothes pole and shelf.

**The Basement.** Space for heating plant, laundry and storage.

**Height of Ceilings.** First floor, 9 feet from floor to ceiling. Second floor, 8 feet 6 inches from floor to ceiling. Basement, 7 feet from floor to joists.

### What Our Price Includes

At the price quoted we will furnish all the material to build this eight-room house, consisting of:

**Lumber; Lath;**
**Roofing,** Oriental Asphalt Shingles, 17-Year Guarantee;
**Framing Lumber,** No. 1 Quality Douglas Fir or Pacific Coast Hemlock;
**Flooring,** Clear Maple for Kitchen and Bathroom, Clear Oak for First Floor Except Kitchen, Clear Douglas Fir or Pacific Coast Hemlock for Second Floor Except Bath.
**Finishing Lumber;**
**High Grade Millwork** (see pages 110 and 111);
**Interior Doors,** One-Panel Fir;
**Trim,** Beautiful Grain Douglas Fir or Yellow Pine;
**Windows,** California Clear White Pine;
**Medicine Case; Bookcases;**
**Built-In Ironing Board; Seat;**
**Mantel;**
**Eaves Trough and Down Spout;**
**40-Lb. Building Paper; Sash Weights; Metal Lath;**
**Chicago Design Hardware** (see page 132);
**Paint** for Three Coats Outside Trim;
**Shellac and Varnish** for Interior Trim and Doors;
**Shellac, Paste Filler and Floor Varnish** for Oak and Maple Floors.

Complete Plans and Specifications.

We guarantee enough material to build this house. Price does not include cement, brick or plaster. See description of "Honor Bilt" Houses on pages 12 and 13.

### OPTIONS

*Sheet Plaster and Plaster Finish, to take the place of wood lath and plaster, $255.00 extra. See page 109.*

*Oak Doors and Trim, for living room, dining room and sun room, also oak stairs, $219.00 extra.*

*Storm Doors and Windows, $112.00 extra.*

*Screen Doors and Windows, galvanized wire, $69.00 extra.*

For prices of Plumbing, Heating, Wiring, Electric Fixtures and Shades, see pages 130 and 131.

A bedroom that readily wins approval because of good size and sunshine.

The modern housewife finds her daily work a pleasurable occupation in the Alhambra kitchen.

# The ALHAMBRA INTERIORS

The above illustration shows the exterior of The Alhambra "Honor Bilt" Modern Home. It is a gem that wins instant admiration. Those who live in it are justly proud.

At the left is a most comfortable and attractive living room. The fireplace grouping is the center of interest for the family and friends.

*Above*—A charming sun room where cheerfulness and restfulness are always present.

*Above*—The dining room reflects formality and harmony. It is a bright and interesting room.

THE GLEN FALLS home literally reproduces the picturesque quality of colonial architecture, introduced by Dutch settlers in the neighborhood of New York.

A motor tour through some fine settlements, including Long Island, will reveal old Dutch farmhouses, fortunately preserved by succeeding generations of owners, that are strikingly similar. Yet, each house is delightfully different—individual! Breathing an air of charming simplicity; the quaint land of dykes and windmills; pleasing and exclusive as any refined American suburb.

The architecture brought over the Atlantic by Dutch settlers is famous for hospitable and dignified appearance, comfortable and economical arrangement, and strong and permanent construction.

The Glen Falls represents, by far, the choicest design of Dutch colonial architecture. Its popularity is a natural result of widespread appreciation for the "home beautiful." Its distinctive shape lends itself remarkably well to the ideal treatment of carrying the lines down to the ground. Carefully planned to stress the harmonious lines are the colonial entrance door with its brick terrace and bench, its set-in porch with dignified wood columns, green shutters and flower boxes. Then, again, con-

**Honor Bilt**

## The Glen Falls
### No. P3245 "Already Cut" and Fitted
### $4,909.00

sider the wide siding, stuccoed fireplace chimney, topped with pots through which the smoke rises from a crackling fire below.

Really, words are of no avail in describing the beauty of The Glen Falls Home. One must study the illustration on this page and the floor plans on the opposite page to fully appreciate the splendid value made possible by our ready-cut system and direct-from-factory-to-you plan.

The plan as illustrated is 46 feet wide by 32 feet deep and will show to the best advantage when placed on a lot 60 feet or more in width. It should be set back a considerable distance from the established walk line in order to achieve a successful perspective from the street or road.

### FIRST FLOOR

**The Reception Hall.** Entrance through the colonial front door leads into the vestibule, which connects with the reception hall by a plastered arch. The space at the left of the vestibule, opening off the hall, can be used for a lavatory or for a closet for wraps, if so desired. The reception hall contains a

graceful colonial stairway with curved rail, and starting tread ascends to the second floor hall. The space beneath the stairway is used for another closet. At the rear end of the hall a door connects with a quiet room, suitable for den, office or studio. It is lighted by a double window. Size, 12 feet by 7 feet 6 inches. The reception hall is 7 feet 6 inches by 19 feet.

**The Living Room.** French doors connect the reception hall with the spacious living room. Size, 21 feet 6 inches by 15 feet. Directly opposite the entrance from the reception hall is a beautiful colonial fireplace and mantel, on either side of which is a pair of casement sash. French doors in pairs open into a sun porch, with smooth cement floor, size, 16 feet by 7 feet, screened and glazed for year round use, and in the front open to the porch shown in the photograph, which has a cement floor in tile effect, and is 9 feet by 22 feet.

The living room offers a splendid opportunity for decoration and furnishing in either colonial period or the more recent mode of individualized pieces, each representing its own origin.

**The Dining Room.** French doors open from the reception hall into the dining room. Floor dimensions are 14 feet 6 inches by 13 feet 3 inches. It is laid out successfully because of the liberal spacing for a complete dining set and comfortable placing of guests. It is lighted by a window in the front and a double window on the side.

**The Dining Alcove.** A swinging door connects the dining room with the dining alcove. It has a five-foot china cabinet, breakfast table with benches, illustrated on pages 110 and 111.

(*Continued on next page*)

(*Continued from the preceding page*)

**The Kitchen.** Size, 14 feet 6 inches by 11 feet 3 inches. Conveniently located is a built-in kitchen cabinet, the kitchen De Luxe Outfit, which includes a sink and built-in cupboards, a broom closet, space for range and table. A double window directly over the sink provides light and ventilation. A door leads to the rear entry, which has space for refrigerator, stairs to basement and door to grade.

## SECOND FLOOR

**The Bedrooms.** The second floor plan contains a large central hall, with built-in window seat in the rear. The hall connects with the four bedrooms, hall bathroom and linen closet. The master's bedroom, in the left front, 19 feet 6 inches by 11 feet 3 inches, easily accommodates twin beds and the usual furniture. It contains a private bathroom, which has a built-in medicine case, and a window. This bedroom has a double window in the front and one window in the side, providing cross ventilation. A clothes closet is another feature. The bedroom in the rear of the master's bedroom has a very attractive built in wardrobe, illustrated on page 111. The bedroom is lighted by two windows. Each of the two bedrooms on the right side of plan has cross ventilation and a spacious clothes closet.

**The Bathroom** in the hall contains a built-in medicine case and a window.

**The Basement.** Excavated under all main parts of the house. Room for heating plant, storage, fruit room and laundry.

**Height of Ceilings.** First floor, 8 feet 6 inches from floor to ceiling. Second floor,

8 feet from floor to ceiling. Basement, 7 feet from floor to joists.

### *What Our Price Includes*

**At the price quoted we will furnish all the material to build this nine-room two-story house, consisting of:**

**Lumber; Lath;**

**Roofing,** Best Grade Clear Red Cedar Shingles;

**Siding,** Clear Cypress or Clear Red Cedar Bevel;

**Framing Lumber,** No. 1 Quality Douglas Fir or Pacific Coast Hemlock;

**Flooring,** Clear Oak for All Rooms except Kitchen and Bath; Kitchen, Clear Maple; Bathrooms, Tile;

**Porch Ceiling,** Clear Grade Douglas Fir or Pacific Coast Hemlock;

**Finishing Lumber;**

**High Grade Millwork** (see pages 110 and 111);

**Interior Doors,** Two Panel Design of Douglas Fir;

**Trim,** Beautiful Grain Douglas Fir or Yellow Pine;

**Windows** of California Clear White Pine;
**Medicine Case; Kitchen Cabinets;**

**China Cabinet;**

**Wardrobe;**

**Porch Seat; Built-in Hall Seat;**

**Mantel; Coal Chute;**

**Kitchen De Luxe Outfit;**

**Breakfast Alcove Table and Seats;**

**Colonial Shutters;**

**Eaves Trough and Down Spout;**

**40-Lb. Building Paper;**

**Sash Weights;**

**Chicago Design Hardware** (page 132);

**Paint** for Three Coats Outside Trim and Siding;

**Enamel** for All Interior Trim and Woodwork.

**Stain and Varnish** for Interior Doors, Stair Treads and Stair Railing.

**Varnish and Wood Filler** for Oak and Maple Floors.

We guarantee enough material to build this house. Price does not include cement, brick or plaster.

See description of "Honor Bilt" Houses on pages 12 and 13.

### OPTIONS

*Sheet Plaster and Plaster Finish,* to take the place of wood lath, $349.00 extra. (See page 109.)

*Storm Doors and Windows,* $162.00 extra.

*Screen Doors and Windows,* galvanized wire, $135.00 extra.

*Oriental Slate Surfaced Shingles in place of wood shingles,* $64.00 extra.

*Birch Trim and Doors for living room, dining room, den, vestibule and hall on first floor and main stairs of birch,* $225.00 extra.

For prices of Plumbing, Heating, Wiring, Electric Fixtures and Shades see pages 130 and 131.

**See Illustration of This Fine Home on Opposite Page.**

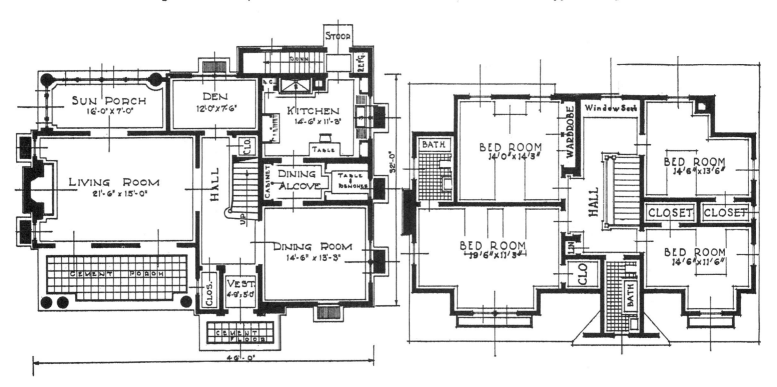

FIRST FLOOR PLAN          SECOND FLOOR PLAN

*For Our Easy Payment Plan See Page 144*

## The Clyde

### No. P9030A "Already Cut" and Fitted
### $1,666.00

THE CLYDE has been built in many sections of the country by customers who tell of their satisfaction. Their letters praise our "Honor Bilt" system, the quality of lumber and millwork. Some tell of the money saved and our reliable service. Others write of having sold their home at a profit and ordering another Sears, Roebuck and Co. "Honor Bilt" home.

Wood Shingle panels and tapered columns, brackets and other little touches make The Clyde an unusually well balanced and attractive house which will look as well on a narrow lot as on a wide one.

**The Living Room.** Entering the living room, it is surprising to find how large and light it is. It extends entirely across the house and has three windows besides the handsome glass door and a sash on each side of the brick mantel. A mantel and fireplace occupy the center of the left wall with a sash on each side. The size of living room is 19 feet 2 inches by 11 feet 2 inches. It is of such good shape and arrangement that no matter what articles of furniture you have, they can always be comfortably and attractively placed.

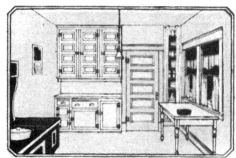

Kitchen showing cabinets we furnish with No. P9030A

**The Dining Room.** The wide cased opening into the dining room adds to the light and the feeling of spaciousness. Here you have the two windows pouring light right over the table. Opposite is a big wall space that will accommodate a buffet.

**The Kitchen.** From the dining room you enter the kitchen, which is an unusually compact, convenient workroom. Standing at the sink you are close to the window and the table. Kitchen cabinet is near range space. The grade entrance is a modern improvement that you will like better the longer you live in the house. It has space for ice box. Fine for carrying things between basement and yard and makes the kitchen easier to heat.

**The Bedrooms.** A hall is open from the dining room that connects with the two bedrooms and bath. There is a clothes closet off each bedroom. The bathroom has a medicine case and is lighted by a window.

**Basement.** Excavated basement under the entire house. Room for storage, furnace and laundry.

**Height of Ceilings.** Basement, 7 feet from floor to joists. Main floor, 9 feet from floor to ceiling.

### What Our Price Includes

**At the price quoted we will furnish all the material to build this five-room bungalow consisting of:**

**Lumber; Lath;**
**Roofing,** Best Grade Clear Red Cedar Shingles;
**Siding,** Clear Cypress or Clear Red Cedar, Bevel;
**Framing Lumber,** No. 1 Quality Douglas Fir or Pacific Coast Hemlock;
**Flooring,** Clear Grade Douglas Fir or Pacific Coast Hemlock;
**Porch Flooring,** Clear Edge Grain Fir;
**Porch Ceiling,** Clear Grade Douglas Fir or Pacific Coast Hemlock;
**Finishing Lumber;**
**Windows** of California Clear White Pine;
**High Grade Millwork** (see pages 110 and 111);
**Interior Doors,** Five-Cross Panel Design Douglas Fir;
**Trim,** Beautiful Grain Douglas Fir or Yellow Pine;
**Mantel; Medicine Case; Kitchen Cabinet;**
**40-Lb. Building Paper; Sash Weights;**
**Eaves Troughs; Down Spout;**
**Stratford Design Hardware** (see page 132);
**Paint for Three Coats Outside Trim and Siding;**
**Shellac and Varnish** for Interior Trim and Doors;
**Stain for Two Brush Coats for Wood Shingles in Gables;**

— Complete Plans and Specifications.

Built on a concrete foundation and excavated under entire house.

We guarantee enough material to build this house. Price does not include cement, brick or plaster.

See description of "Honor Bilt" Houses on pages 12 and 13.

FLOOR PLAN

Can be built on a lot 28 feet wide

### OPTIONS

*Sheet Plaster and Plaster Finish, instead of wood lath, $132.00 extra. See page 109.*

*Oriental Asphalt Shingles, guaranteed 17 years, instead of wood shingles, $33.00 extra.*

*Oak Doors, Trim and Floors in living room and dining room. Maple Floors in kitchen and bathroom, $128.00 extra.*

*Storm Doors and Windows, $53.00 extra.*

*Screen Doors and Windows, galvanized wire, $35.00 extra.*

For prices of Plumbing, Heating, Wiring, Electric Fixtures and Shades see pages 130 and 131.

### For Our Easy Payment Plan See Page 144

THE OLIVIA bungalow surely deserves, its great popularity because it is one of the best planned four-room-with-bath bungalows. Here the architect has created a beautiful and harmonious design and, in addition, has planned the greatest amount of available floor space without wasting one inch.

Observe the beautiful front porch with large gable roof with exposed rafters and fancy verge boards. Note the artistic arrangement of concrete columns and how the center one provides a convenient place for a jardiniere or flower box. Then, note the shingled gables and paneled columns. Follow the lines of this home from front to rear and there will be no doubt in your mind that this is a most attractive home. Gray painted trim with a white color body will make this the "niftiest" house in your block.

The front porch, 16 feet by 7 feet, is delightful. It may be screened in summer and glazed in winter. A swing or a lounging chair, with table, lamp and rug, and the porch is converted into a sun room.

**The Living Room.** Size, 10 feet 8 inches by 15 feet 2 inches. Long wall spaces permit the placing of furniture and piano in a pleasing manner. There is plenty of light and ventilation from two windows and glazed front door.

**The Kitchen and Dining Room.** From the living room a door opens into the large kitchen. It is 8 feet 7 inches by 12 feet 8 inches. Location of sink and stove are planned to save many steps when preparing the daily meals. On the opposite side there is a built-in cabinet. (See illustration to the left.) Near the stove is a cased opening to the pantry. It is provided with shelves for

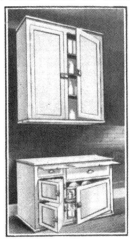

Cabinets Furnished in Kitchen

**Honor Bilt**

# The Olivia
### No. P7028 "Already Cut" and Fitted
## $1,283.00

utensils and other kitchen needs. A table can be placed under the two windows. Directly off the kitchen is a large closet with two shelves. A door leads down three steps to the rear entry, where there is space for an ice box. Steps to basement are here, also.

**The Bedrooms.** From the living room a hall connects with the two bedrooms and bath. Front bedroom is 10 feet 2 inches by 10 feet 8 inches, and rear bedroom is 10 feet 2 inches by 9 feet 7 inches. Each bedroom has a clothes closet. There are two windows in each bedroom, permitting cross ventilation and light.

The bathroom plumbing is arranged on one wall, saving material and labor. The bathroom has a medicine cabinet and a window.

**Basement.** Excavated basement with concrete floor. Room for furnace, laundry and storage.

**Height of Ceilings.** Main floor, 8 feet 2 inches from floor to ceiling. Basement, 7 feet from floor to joists.

### What Our Price Includes

At the price quoted we will furnish all the material to build this four-room bungalow, consisting of:
Lumber; Lath;
Roofing, Best Grade Clear Red Cedar Shingles;
Siding; Clear Grade Cypress or Clear Red Cedar, Bevel;
Framing Lumber, No. 1 Quality of Douglas Fir or Pacific Coast Hemlock;
Flooring, Clear Grade Douglas Fir or Pacific Coast Hemlock;
Porch Ceiling, Clear Grade Douglas Fir or Pacific Coast Hemlock;
Finishing Lumber;
High Grade Millwork (see pages 110 and 111);
Interior Doors, Five-Cross Panel Design of Douglas Fir;
Trim, Beautiful Douglas Grain Fir or Yellow Pine;
Windows of California Clear White Pine;
Medicine Case;
Kitchen Cabinet;
Eaves Trough and Down Spouts;
40-Lb. Building Paper; Sash Weights;
Stratford Design Hardware (see page 132);
Paint for Three Coats Outside Trim and Siding;
Shellac and Varnish for Interior Doors and Trim.

Complete Plans and Specifications.

We guarantee enough material to build this house. Price does not include cement, brick or plaster. See description of "Honor Bilt" Houses on pages 12 and 13.

This house can be built with rooms reversed. See page 3.

**Can be built on a lot 28 feet wide.**

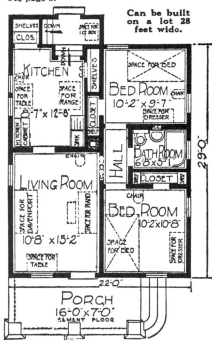

FLOOR PLAN

### OPTIONS

*Sheet Plaster and Plaster Finish, to take the place of wood lath, $124.00 extra. See page 109.*

*Oriental Asphalt Shingles, guaranteed 17 years, instead of wood shingles, $34.00 extra.*

*Storm Doors and Windows, $41.00 extra.*

*Screen Doors and Windows, galvanized wire, $28.00 extra.*

For prices of Plumbing, Heating, Wiring, Electric Fixtures and Shades see pages 130 and 131.

*For Our Easy Payment Plan See Page 144*

THE VALLONIA is a prize bungalow home. It has been built in hundreds of localities. Photographs and many testimonials confirm the splendid features and value. Customers tell of saving as much as $2,500.00 on their Vallonia and often selling at a big profit.

The Vallonia is favored by a sloping, overhanging roof and shingled dormer which has three windows. Roof has a timber cornice effect. Sided with cypress (the wood eternal). Porch extends entirely across front of house, with latticework beneath porch floor. Here under the shade of its roof (with option of screening part or entire porch), you may enjoy leisure hours during summer on a swing or easy chair. Children quickly adopt this porch for play or study.

Perfect harmony in all details marks the architecture of the Vallonia. Every inch of material is "Honor Bilt" quality and workmanship.

### FIRST FLOOR

**The Living Room.** A handsome San Jose door opens into the cheerful living room. Size, 15 feet 2 inches wide by 12 feet 1 inch deep. Liberal wall space accommodates piano and furniture. Three windows provide light and air.

**The Dining Room.** A cased opening divides living and dining rooms. Here the usual dining furniture has enough space for an attractive arrangement. Two windows assure light and ventilation.

**The Kitchen.** From the dining room a swinging door opens into the pleasant kitchen. The range space is just inside the door. Beneath the double window is plenty of space for a table and chairs. Sink is underneath the rear window. A door opens to a convenient pantry, lighted by a window. Everything about this kitchen has been planned to win the lasting approval of the housewife. Just outside the door is the enclosed rear entry with space for refrigerator. Here, also, are stairs to grade and basement.

**The Bedrooms.** From dining room a passage opens into hall that gives privacy to front and rear bedrooms and bath. There is a clothes closet in the passage and a linen closet in the hall. Each bedroom has two windows and a clothes closet. A feature of the rear bedroom is a storage closet with shelves. The bathroom has a medicine case and a window.

### SECOND FLOOR

An enclosed stairway leads from the dining room to second floor. We will furnish material for three bedrooms and three closets on second floor, with single floor for $247.00 extra. See plan at lower right.

**Basement.** Room for furnace, laundry and storage.

**Height of Ceilings.** Basement, 7 feet from floor to joists. Main floor, 9 feet from floor to ceiling.

---

*The Vallonia Is Shown in Colors on Page 20*

**Honor Bilt**

## The Vallonia
### No. P13049A "Already Cut" and Fitted
### $2,076.00

### What Our Price Includes

At the price quoted we will furnish all the material to build this five-room bungalow, consisting of:
Lumber; Lath;
**Roofing,** Best Grade Clear Red Cedar Shingles;
**Siding,** Clear Cypress or Clear Red Cedar, Bevel;
**Framing Lumber,** No. 1 Quality Douglas Fir or Pacific Coast Hemlock;
**Flooring,** Clear Grade Douglas Fir or Pacific Coast Hemlock;
**Porch Flooring,** Clear Edge Grain Fir;
**Porch Ceiling,** Clear Grade Douglas Fir or Pacific Coast Hemlock;
**Finishing Lumber;**
**High Grade Millwork** (see pages 110 and 111);
**Interior Doors,** Two Inverted Cross Panel Design of Douglas Fir;
**Trim,** Beautiful Grain Douglas Fir or Yellow Pine;
**Medicine Case;**
**Windows** of California Clear White Pine;
**40-Lb. Building Paper; Sash Weights;**
**Eaves Trough; Down Spout;**
**Chicago Design Hardware** (see page 132);
**Paint** for Three Coats Outside Trim and Siding;
**Stain** for Shingles on Dormer Wall for Two Brush Coats;
**Shellac and Varnish** for Interior Trim and Doors.
Complete Plans and Specifications.
Built on concrete foundation and excavated under entire house.
We guarantee enough material to build this five-room bungalow. Price does not include cement, brick and plaster. "Honor Bilt" Construction explained on pages 12 and 13.

### OPTIONS

*Sheet Plaster and Plaster Finish, to take the place of wood lath, $172.00 extra, for first floor; with attic, $269.00 extra. See page 109.*

*Oak Doors, Trim and Floors for living room and dining room, Maple Floors in kitchen and bathroom, $145.00 extra.*

*Oriental Slate Surfaced Asphalt Shingles, instead of wood shingles, $46.00 extra.*

*Storm Doors and Windows, $54.00 extra, without attic, and $70.00 extra, with attic.*

*Screen Doors and Windows, galvanized wire, $35.00 extra, without attic, and $46.00 extra, with attic.*

For Prices of Plumbing, Heating, Wiring, Electric Fixtures and Shades see pages 130 and 131.

*For Our Easy Payment Plan See Page 144*

---

**Can be built on a lot 34 feet wide**

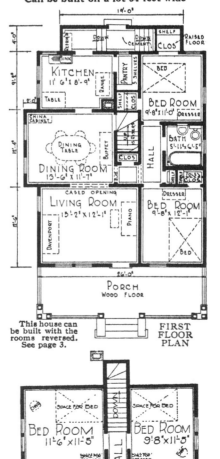

This house can be built with the rooms reversed. See page 3.

**FIRST FLOOR PLAN**

Finished Second Floor Plan $247.00 Extra

---

# The VALLONIA INTERIORS

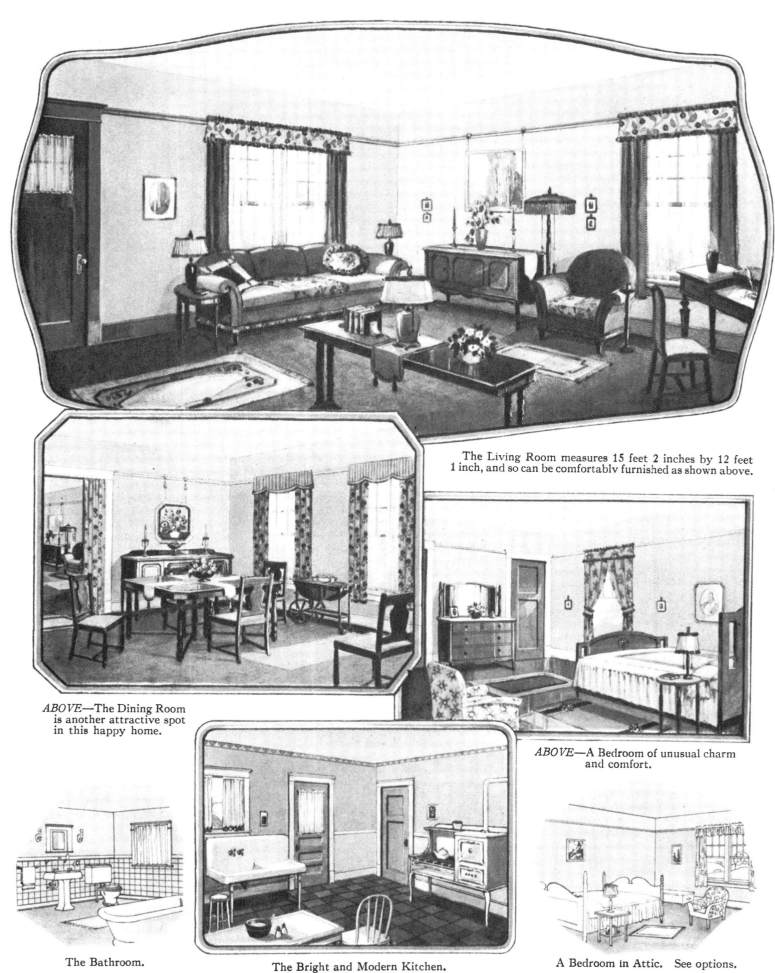

The Living Room measures 15 feet 2 inches by 12 feet 1 inch, and so can be comfortably furnished as shown above.

*ABOVE*—The Dining Room is another attractive spot in this happy home.

*ABOVE*—A Bedroom of unusual charm and comfort.

The Bathroom.

The Bright and Modern Kitchen.

A Bedroom in Attic.   See options.

**Can be built on a lot 30 feet wide**

FIRST
FLOOR PLAN

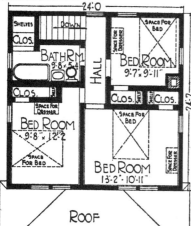

SECOND FLOOR PLAN

THIS popular style of architecture provides for the greatest amount of comfort in a two-story six-room house. The simplicity of the exterior, together with the fact that there is no waste space, makes the Gladstone a high grade house at a low price.

Shingles on the roof, dormer and porch can be stained either bungalow brown or moss green to harmonize with the painted cypress siding. Porch is 8 feet wide by 24 feet long. Foundation for porch can be built of brick or similar material, with a concrete cap serving to support the twin columns. If screened or glazed the porch will make a comfortable room in season. The house is 24 feet wide by 24 feet 7 inches long.

### FIRST FLOOR

**The Living Room.** Entering the house from the porch you pass through a bevel plate glass door. You will be charmed with the living room, which is 11 feet 5 inches wide by 16 feet 5 inches long. Here ample space is provided for piano, library table, davenport and other furniture. Full length plate glass mirror door opens into a clothes closet equipped with wardrobe pole and shelf. From the end of this room a stairway leads to the second floor. Living room is well lighted and cross ventilated by three windows.

**The Dining Room.** A cased opening between living room and dining room makes them available as one very large room when entertaining. Dining room is 11 feet 5 inches wide by 13 feet 7 inches long, and contains space to seat the family and friends. An abundance of sunshine and air is secured by the double window in front and a window at the side.

**The Kitchen.** The kitchen is directly off the dining room and has a swinging door. A table may be placed beneath the double window. The location provides for cupboard, table, sink and range, reduces steps and saves time. Air and light come from three windows.

The pantry with shelves and space for refrigerator, well lighted by window, opens from the kitchen. Here steps lead to grade entry opening to yard and to basement.

### SECOND FLOOR

**The Bedrooms.** Three bedrooms and bath are entered from hall on second floor. Ample sunshine and air are furnished the front bedrooms through double windows and one side window in each room. The rear bedroom has two windows. A clothes closet with wardrobe rod and shelf is provided for each bedroom.

**The Bathroom.** All plumbing is on one wall, affording installation economy. Wise use of space is demonstrated by a linen closet in this room.

**The Basement.** Basement under entire house is lighted by four hinge sash. Room for furnace, laundry and storage.

**Height of Ceilings.** Basement, 7 feet from concrete floor to joists. Main floor, 9 feet 2 inches from floor to ceiling. Second floor, 8 feet 8 inches from floor to ceiling.

### The Gladstone
#### No. P3222 "Already Cut" and Fitted
## $2,029.00

### What Our Price Includes

At the price quoted we will furnish all the material to build this six-room house, consisting of:
**Lumber; Lath;**
**Roofing,** Best Grade Clear Red Cedar Shingles;
**Siding,** Clear Cypress or Clear Red Cedar, Bevel;
**Framing Lumber,** No. 1 Quality Douglas Fir or Pacific Coast Hemlock;
**Flooring,** Clear Douglas Fir or Pacific Coast Hemlock;
**Porch Flooring,** Clear Edge Grain Fir;
**Porch Ceiling,** Clear Douglas Fir or Pacific Coast Hemlock;
**Finishing Lumber;**
**High Grade Millwork** (see pages 110 and 111);
**Interior Doors,** Two Vertical Panel Design of Douglas Fir;
**Trim,** Beautiful Grain Douglas Fir or Yellow Pine;
**Medicine Case;**
**Windows** of California Clear White Pine;
**40-Lb. Building Paper; Sash Weights;**
**Eaves Troughs; Down Spout;**
**Chicago Design Hardware** (see page 132);
**Paint** for Three Coats Outside Trim and Siding;
**Shellac and Varnish** for Interior Trim and Doors.
Complete Plans and Specifications.

Built on concrete foundation and excavated under entire house.

We guarantee enough material to build this house. Price does not include cement, brick or plaster.

See description of "Honor Bilt" Houses on pages 12 and 13.

### OPTIONS

*Sheet Plaster and Plaster Finish, to take the place of wood lath, $208.00 extra. See page 109.*

*Oriental Asphalt Shingles, guaranteed 17 years, instead of wood shingles for roof, $29.00 extra.*

*Oak Doors, Trim and Floors in living room and dining room, also Oak Stairs, Maple Floors in kitchen and bathroom, $162.00 extra.*

*Storm Doors and Windows, $74.00 extra.*

*Screen Doors and Windows, galvanized wire, $49.00 extra.*

For Prices of Plumbing, Heating, Wiring, Electric Fixtures and Shades see pages 130-131.

*For Our Easy Payment Plan See Page 144*

# The GLADSTONE INTERIORS

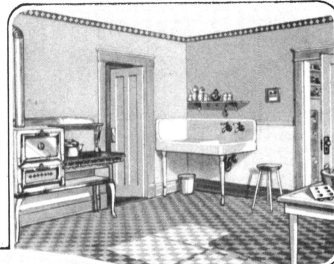

*Above*—The dining room atmosphere is cordial and vivid.

*Center*—Living room where one may entertain his friends confident of their hearty approval.

All of the furnishings shown in these interiors were selected from the pages of our big General Catalog, "The Thrift Book of a Nation."

*Above*—Sunshine floods this example of modern kitchen efficiency.

*At left*—The simple beauty of the bedroom tells its own story.

"Happiness abides with folks who recognize the value of a good home for themselves and their children."

—CHARLOTTE DEXTER

## The Homewood

### No. P3238 "Already Cut" and Fitted
### $2,809.00

THE HOMEWOOD has the cozy appearance so much admired in the smaller houses, yet skillful planning enables it to retain the good size and decided advantage of the large two-story homes. Considerable thought has been given to the design, size, style of the windows, fireplace, chimney, shutters, front porch, trellises and shingles. Moreover, the interior has been planned along modern lines to afford the greatest livable space, convenient arrangement and imparting an air of quality.

#### FIRST FLOOR

**The Porch,** size, 18 feet by 8 feet, is constructed with solid rail and heavy shingle columns which, together with the careful handling of the trellis work, gives it a semi-private exposure.

**The Reception Hall** has a semi-open stairway to the second floor, a closet for wraps, and a pair of French doors that lead to the living room.

**The Living Room** dimensions are 19 feet 5 inches by 13 feet 5 inches. In the center of the right wall is a mantel and fireplace. On each side of the fireplace is a high sash window with room underneath for bookcases or other furniture. Good spacing accommodates a piano and appropriate furniture. In the front are two windows providing plenty of light and air.

**The Dining Room** is square in shape, 13 feet 5 inches by 13 feet 5 inches. There is a triple high sash in the side and two windows in the rear, making a pleasant, cheery and well ventilated room. A buffet can be placed underneath the high sash. Serving is very convenient because of the good spacing.

**The Kitchen.** A swinging door connects the dining room with the kitchen. Here, in the housewife's workshop, is a built-in cupboard on each side of the sink, opposite which is space for range and a broom closet. A double window directly over the sink and another window on the side wall provide plenty of light and cross ventilation. A door connects with the side entry, which has space for a refrigerator, stairs to basement and grade.

#### SECOND FLOOR

**Stairway** from the reception hall leads to the hall on the second floor, which connects with the three bedrooms, bathroom and linen closet. The main feature of the second floor is the size of the master's bedroom. It is 13 feet 5 inches by 16 feet 11 inches and will accommodate twin beds, besides other pieces of furniture. A double window in the front and a side window admits light and cross ventilation. It has a clothes closet. Each of the other two bedrooms has a double window, the front bedroom has one clothes closet, while the rear bedroom has a clothes closet and another closet for storing the off-season garments.

**The Bathroom** has a built-in medicine case and is lighted by a window.

**The Basement.** Room for furnace, laundry and storage.

**Height of Ceilings.** First floor, 8 feet 6 inches from floor to ceiling. Second floor, 8 feet 6 inches from floor to ceiling. Basement, 7 feet from floor to joists.

### What Our Price Includes

At the price quoted we will furnish all the material to build this six-room, two-story bungalow, consisting of:

Lumber; Lath;
Roofing, Best Grade Clear Red Cedar Shingles;
Siding, Best Grade Clear Red Cedar Shingles;
Framing Lumber, No. 1 Quality Douglas Fir or Pacific Coast Hemlock;
Flooring, Clear Grade Douglas Fir or Pacific Coast Hemlock;
Porch Flooring, Clear Grade Douglas Fir or Pacific Coast Hemlock;
Porch Ceiling, Clear Grade Douglas Fir or Pacific Coast Hemlock;
Finishing Lumber;
High Grade Millwork (see pages 110 and 111);
Interior Doors, Two Panel Design of Douglas Fir;
Trim, Beautiful Grain Douglas Fir or Yellow Pine;
Windows of California Clear White Pine;
Medicine Case;
Kitchen De Luxe Outfit;
Mantel; Colonial Shutters;
Eaves Trough and Down Spout;
40-Lb. Building Paper; Sash Weights;
Chicago Design Hardware (see page 132);
Paint for Three Coats Outside Trim;
Stain for Shingles on Walls for Two Brush Coats;
Varnish and Shellac for Interior Doors and Trim.

Complete Plans and Specifications.

We guarantee enough material to build this house.

Price does not include cement, brick or plaster.

See description of "Honor Bilt" houses on pages 12 and 13 of Modern Homes Catalog.

### OPTIONS

*Sheet Plaster and Plaster Finish, to take the place of wood lath, $231.00 extra. See page 109.*

*Storm Doors and Windows, $108.00 extra.*

*Screen Doors and Windows, galvanized wire, $70.00 extra.*

*Oriental Slate Surfaced Shingles, in place of wood shingles, for roof, $36.00 extra.*

For prices of Plumbing, Heating, Wiring, Electric Fixtures and Shades see pages 130 and 131.

Can be built on a lot 35 feet wide

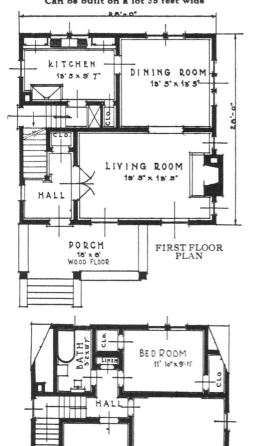

FIRST FLOOR PLAN

SECOND FLOOR PLAN

### For Our Easy Payment Plan See Page 144

THE SOLACE is a one and a half story bungalow type with a dormer on the rear. The exterior and interior have been carefully planned. Its six rooms and bath reflect modern comforts. In the Solace there is a great economy of floor space and low cost of upkeep. Every dollar's worth of material shows up to the best advantage. It has the same high grade material as found in the higher priced houses. The porch has a covered roof with pergola effect. This lends character and individuality.

### FIRST FLOOR

**The Living Room.** Entering the living room from the front porch you will observe the nice living room with ample wall space for piano and furniture. The living room has a large cased opening into dining room, which is used as one large room when desired.

**The Dining Room.** The dining room has a space for a buffet. Light, sunshine and air from double windows on the side and another window in the rear.

**The Kitchen.** From the dining room a door leads to the modern kitchen. The sink, the range space, the kitchen cabinet with its labor saving devices, and space for ice box under a part of the cabinet are all grouped to save work for the busy housewife. Light comes from two windows, one in the rear and the other on the side.

**The Bedroom.** The front bedroom and bath open from a hall that connects with all rooms. The bedroom on first floor has two windows and a clothes closet. The bathroom is just the convenient size. It has a medicine case.

### SECOND FLOOR

**The Bedrooms.** Directly off the hall on the first floor is the closed stairway with door that leads to the second floor. The hall is lighted by a window in the dormer. Two large bedrooms open from the hall. Bedrooms have double windows, giving plenty of sunshine and air. A feature worth mentioning is the large storage or closet space available to each bedroom through a door.

**Basement.** Excavated under the entire house.

**Height of Ceilings.** Basement, 7 feet from floor to ceiling. First floor, 8 feet 6 inches from floor to ceiling. Second floor, 8 feet 6 inches from floor to ceiling.

---

**Honor Bilt**

## The Solace
*No. P3218 "Already Cut" and Fitted*
### $1,579.00

#### *What Our Price Includes*
At the price quoted we will furnish all the material to build this six-room bungalow, consisting of:
Lumber; Lath;
Roofing, Best Grade Clear Red Cedar Shingles;
Siding, Clear Cypress or Clear Red Cedar, Bevel;
Framing Lumber, No. 1 Quality Douglas Fir or Pacific Coast Hemlock;
Flooring, Clear Douglas Fir or Pacific Coast Hemlock;
Porch Flooring, Clear Edge Grain Fir;
Porch Ceiling, Clear Grade Douglas Fir or Pacific Coast Hemlock;
Finishing Lumber;
High Grade Millwork (see pages 110 and 111);
Interior Doors, Two Cross Panel Design of Douglas Fir;
Trim, Beautiful Grain Douglas Fir or Yellow Pine;
Kitchen Cabinet; Medicine Case;
Windows of California Clear White Pine;
Eaves Trough; Down Spout;
40-Lb. Building Paper; Sash Weights;
Stratford Design Hardware (see page 132);
Paint for Three Coats of Outside Trim and Siding;
Shellac and Varnish for Interior Trim and Doors.
Complete Plans and Specifications.
Built on a concrete foundation, brick above grade.
We guarantee enough material to build this house. Price does not include cement, brick or plaster.
See description of "Honor Bilt" Houses on pages 12 and 13.

#### OPTIONS
*Sheet Plaster and Plaster Finish, to take the place of wood lath and plaster, $164.00 extra. See page 109.*
*Storm Doors and Windows, $63.00 extra.*
*Screen Doors and Windows, galvanized wire, $41.00 extra.*
*Oriental Slate Surfaced Shingles, guaranteed 17 years, instead of wood shingles for roof, $36.00 extra.*
For prices of Plumbing, Heating, Wiring, Electric Fixtures and Shades see pages 130 and 131.

**Rear View**

---

Can be built on a lot 30 feet wide

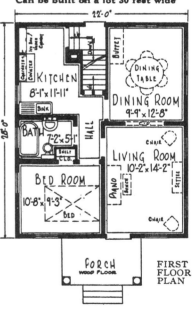

FIRST FLOOR PLAN

This house can be built with rooms reversed. See page 3.

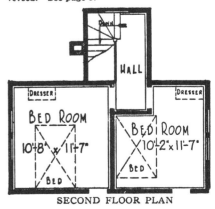

SECOND FLOOR PLAN

---

*For Our Easy Payment Plan See Page 144*

THE WALTON embodies strength, dignity and gracefulness. It presents a most pleasing appearance and is of a character that will long retain popular favor.

**The Living Room.** The six rooms and bath are well arranged. It is not necessary to enter any of the three bedrooms directly from the main rooms. Ample closets are provided and many of the latest ideas for space utilizers are embodied. The living room is of excellent proportions and well lighted. Size, 14 feet 2 inches by 19 feet 1 inch. The fireplace and bookcase colonnade add to comfort as well as beauty.

**The Dining Room.** Through the bookcase colonnade you enter the dining room, 16 feet 2 inches by 12 feet 9 inches. The dining room is flooded with sunshine by the row of windows on the left and with the long wall space opposite these windows you can place a buffet of any size.

**The Kitchen.** The kitchen is planned to make the work easy. It measures 8 feet 9 inches by 12 feet 1 inch. Only a few steps away from the door to the dining room is the stove. On the landing at the grade entrance, just inside the back door which is glazed, is a space for refrigerator. Here, too, are stairs to basement.

**The Bedrooms.** A hallway, which has a coat closet, and a linen closet, connects the living room with two bedrooms and bath.

The front bedroom has a built-in cedar chest that forms a window seat. There is a clothes closet with hat shelf. Light and air come from three windows. The other bedroom has a clothes closet with hat shelf and a double window.

The rear bedroom is entered from a square hall that is open from the dining room. This hall accommodates a clothes closet and stairs to attic. All bedrooms allow comfortable placing of the usual furniture.

A bathroom is between the two bedrooms. The above mentioned linen closet has a door opening into the bathroom.

### Honor Bilt
### The Walton
**No. P13050 "Already Cut" and Fitted**
### $2,471.00

**The Basement.** Excavated basement with concrete floor. Room for furnace, laundry and storage.

**Height of Ceilings.** Basement, 7 feet from floor to joists. Main floor, 9 feet from floor to ceiling.

#### What Our Price Includes

At the price quoted we will furnish all the material to build this six-room bungalow, consisting of:

**Lumber; Lath;**
**Roofing,** Best Grade Clear Red Cedar Shingles;
**Siding,** Clear Cypress or Clear Red Cedar, Bevel;
**Framing Lumber,** No. 1 Quality Douglas Fir or Pacific Coast Hemlock;
**Flooring,** Clear Maple for Kitchen and Bathroom, Clear Oak for Balance of Rooms;
**Porch Floor** to Be Cement;
**Porch Ceiling,** Clear Grade Douglas Fir or Pacific Coast Hemlock;
**Finishing Lumber;**
**High Grade Millwork** (see pages 110 and 111);
**Interior Doors,** Inverted Two-Cross Panel Design of Douglas Fir;
**Trim,** Beautiful Grain Douglas Fir or Yellow Pine;
**Windows** of California Clear White Pine;
**Bookcase Colonnade; Medicine Case;**
**Mantel; Sash Weights; Cedar Chest Seat;**
**40-Lb. Building Paper;**
**Eaves Trough; Down Spouts;**
**Chicago Design Hardware** (see page 132);
**Paint** for Three Coats of Outside Trim and Siding;
**Shellac and Varnish** for Interior Trim and Doors;
**Shellac, Paste Filler and Floor Varnish** for Oak and Maple Floors;
**Stain** for Two Brush Coats for Shingles on Gables.

Complete Plans and Specifications.

Built on a concrete foundation.

We guarantee enough material to build this house. Price does not include cement, brick or plaster.

See Description of "Honor Bilt" Houses on Pages 12 and 13.

**Can be built on a lot 40 feet wide**

**FLOOR PLAN**

#### OPTIONS
*Sheet Plaster and Plaster Finish to take the place of wood lath, $200.00 extra. See page 109.*
*Oriental Asphalt Shingles, guaranteed 17 years, instead of wood shingles, $58.00 extra.*
*Oak Trim and Doors for living room and dining room, $159.00 extra.*
*Storm Doors and Windows, $72.00 extra.*
*Screen Doors and Windows, galvanized wire, $42.00 extra.*
For Prices of Plumbing, Heating, Wiring, Electric Fixtures and Shades see pages 130 and 131.

*For Our Easy Payment Plan See Page 144*

# The WALTON ~ INTERIORS

*ABOVE*—A feeling of comfort and good fellowship is present always in this living room with its handsome fireplace. Here, too, a bookcase colonnade between the living room and dining room not only adds a note of beauty, but it also serves to unite both rooms, affording extra space when entertaining.

*CENTER*—This view of the front bedroom reveals the built-in cedar chest that forms a window seat.

*BOTTOM LEFT*—This up to date kitchen includes a step and time saving built-in kitchen cupboard. Well lighted and ventilated by a double window.

*BOTTOM RIGHT*—Good spacing in this dining room allows complete furnishing, including a buffet, with plenty of room for guests.

Above views show the satisfactory way one owner furnished The Walton.

Can be built on a lot 32 feet wide

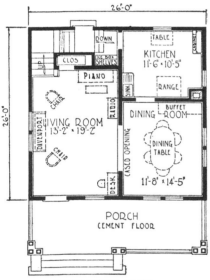

FIRST FLOOR PLAN

SECOND FLOOR PLAN

HERE is a fine two-story home that any American can be proud of and be comfortable in. It is a dignified, substantial house that will stand out among its neighbors and never go "out of style." The rooms of the Americus are all of good size and well lighted and ventilated. Lots of big closets just where needed, and a kitchen that will save a great many steps.

### FIRST FLOOR

**The Porch** is almost a living room during the summer. It may be screened, or if glazed can be used as a sun room. Size, 26 feet by 8 feet.

**The Living Room** is of excellent proportions and has good wall space. It measures 13 feet 2 inches by 19 feet 2 inches. The stairway to the second floor is located directly opposite the front door, near which is a spacious coat closet with a mirror door. Furniture can be grouped to effect almost any kind of arrangement, and there is wall space for a piano. Light and air from windows at the side and front.

**The Dining Room.** A 7-foot cased opening connects the living room and the dining room. Here is room enough for a large buffet, as well as for a china cabinet, serving table or tea wagon. Four big windows allow light and air.

**The Kitchen** connects with the dining room by a swinging door. A couple of steps from the sink is the space for a range, with good light for oven and top. At one corner of the kitchen is a built-in kitchen cabinet, and nearby is space for the table directly underneath a window. Now, if you bake, you have everything together, flour in sanitary swinging bins, all utensils and ingredients, without crossing the room once. At the grade entrance and on the same level with the kitchen floor is space for refrigerator. Over the refrigerator space is a big shelf for things wanted near the kitchen but not used daily. All your work is concentrated on the three walls nearest the dining room. On the other wall is a window and space for table. Notice the grade entrance and the square turn for basement stairs—no winding stairs. Tradesmen and icemen cannot track up your clean kitchen floor, and the kitchen is always warm in winter. There are windows on two sides.

### SECOND FLOOR

**The Bedrooms.** The stairs make a turn and bring you to a small hall from which all rooms and a closet open. A window lights stairs and hall. At the head of the stairs is the bathroom, all plumbing on one wall, and plenty of space for towel rods. At the end of the tub is a generous cupboard, over 5 feet high and 3 feet deep, in the back of which are deep shelves for storage and in the front of which, out of the reach of children, is ample room for towels and a handy shelf for bottles. All three bedrooms are large. Each has windows on two sides and a closet with a shelf. Windows are so arranged that beds can be placed away from them in winter and near them in summer. Off the hall is another closet for bed linen, or storage, or for an extra broom, etc., so that it will not be necessary to carry the cleaning things up and down stairs.

**The Basement.** Room for furnace, laundry and storage. **Height of Ceilings.** First floor, 9 feet from floor to ceiling. Second floor, 8 feet 2 inches from floor to ceiling. Basement, 7 feet from floor to joists.

*For Our Easy Payment Plan See Page 144*

### Honor Bilt
## The Americus
### No. P13063 "Already Cut" and Fitted.
### $2,286.00

#### What Our Price Includes

At the price quoted we will furnish all the material to build this six-room two-story home, consisting of:
**Lumber; Lath;**
**Roofing,** Oriental Asphalt Shingles, Guaranteed 17 Years;
**Siding,** Clear Cypress or Clear Red Cedar, Bevel;
**Framing Lumber,** No. 1 Quality Douglas Fir or Pacific Coast Hemlock;
**Flooring,** First Floor, Clear Oak for Living Room and Dining Room, Clear Maple for Kitchen; Second Floor, Clear Maple for Bathroom, Clear Douglas Fir or Pacific Coast Hemlock for Bedrooms and Hall;
**Porch Ceiling,** Clear Douglas Fir or Pacific Coast Hemlock;
**Finishing Lumber;**
**High Grade Millwork** (see pages 110 and 111);
**Interior Doors,** Two Vertical Panel Design for First Floor and Two-Cross Panel Design for Second Floor, All of Douglas Fir;
**Trim,** Beautiful Grain Douglas Fir or Yellow Pine;
**Windows** of California Clear White Pine;
**Medicine Case;**
**Kitchen Cabinet;**
**Eaves Trough and Down Spout;**
**40-Lb. Building Paper; Sash Weights;**
**Chicago Design Hardware** (see page 132);
**Paint for Three Coats Outside Trim and Siding;**
**Shellac and Varnish** for Interior Doors and Trim;
**Shellac, Paste Filler and Floor Varnish** for Oak and Maple Floors.
Complete Plans and Specifications.
We guarantee enough material to build this house. Price does not include cement, brick or plaster.
See description of "Honor Bilt" Houses on pages 12 and 13.

#### OPTIONS
*Sheet Plaster and Plaster Finish to take the place of wood lath,* $221.00 extra. See page 109.
*Oak Doors and Trim in living room and dining room,* $137.00 extra.
*Wood Floor Material for the porch,* $73.00 extra.
*Storm Doors and Windows,* $78.00 extra.
*Screen Doors and Windows,* galvanized wire, $49.00 extra.
For prices of Plumbing, Heating, Wiring, Electric Fixtures and Shades see pages 130 and 131.

View of end of Living Room, showing stairway and mirror door.

IT IS hardly necessary to say that The Rodessa is a most attractive little home. The illustration proves it beyond a doubt. Furthermore, the price is also attractive. Much thought and expert advice have been employed in designing an exterior that will make this bungalow appeal to lovers of artistic homes. The interior plan appeals to all people desiring the utmost economy in space. The Rodessa has proved to be one of our most popular houses, and owners are delighted with it. If you have only a moderate amount to invest, and wish to secure the biggest value for your money, with the greatest results in comfort, convenience and attractiveness, you can make no mistake in purchasing this house.

**Can be built on a lot 27 feet wide**

This house can be built with the rooms reversed. See page 3

**No. P7041 With Bathroom**

**Honor Bilt**

## The Rodessa
### No. P7041 "Already Cut" and Fitted
### $1,179.00
### No. P3203 Without Bathroom "Already Cut" and Fitted
### $1,182.00

**The Living and Dining Room.** A trellised front porch, 12 feet by 6 feet, leads to the combination living and dining room. The room accommodates a piano, davenport, radio cabinet, upholstered chairs, lamps and a dining set. Size of room, 11 feet 8 inches by 17 feet 8 inches. Sunshine and air are assured by windows at the side and front.

**The Kitchen.** A door connects the living and dining room with the kitchen. Size of kitchen, 8 feet 9 inches by 9 feet 2 inches. There is space for a sink, table, range and cabinet. A double window provides light and ventilation.

Steps in the rear entry lead to the basement.

**The Bedrooms.** From the living and dining room a door opens into the front bedroom. Size of bedroom, 9 feet 2 inches by 9 feet 8 inches. It has a clothes closet with a cased opening, and a front window.

The rear bedroom is entered from a hallway off the living and dining room. It has a clothes closet with a cased opening, and a side window.

**The Bathroom,** too, opens off the hallway. It has a medicine case, and is lighted by a window.

**The Basement.** Room for furnace, laundry and storage.

**Height of Ceilings.** Main floor, 9 feet from floor to ceiling. Basement, 7 feet from floor to joists.

### What Our Price Includes

At the price quoted we will furnish all the material to build this four-room bungalow, consisting of:

Lumber; Lath;
Roofing, Best Grade Clear Red Cedar Shingles;
Siding, Clear Cypress or Clear Red Cedar, Bevel;
Framing Lumber, No. 1 Quality Douglas Fir or Pacific Coast Hemlock;
Flooring, Clear Douglas Fir or Pacific Coast Hemlock;
Porch Flooring, Clear Edge Grain Fir;
Porch Ceiling, Clear Douglas Fir or Pacific Coast Hemlock;
Finishing Lumber;
High Grade Millwork (see pages 110 and 111);
Interior Doors, Five Cross Panel Design of Douglas Fir;
Trim, Beautiful Grain Douglas Fir or Yellow Pine;
Windows, California Clear White Pine;
Medicine Case;
Eaves Trough; Down Spouts;
40-Lb. Building Paper; Sash Weights;
Stratford Design Hardware (see page 132);
Paint for Three Coats Outside Trim and Siding;
Shellac and Varnish for Interior Trim and Doors.
Complete Plans and Specifications.

We guarantee enough material to build this house. Price does not include cement, brick or plaster.

See description of "Honor Bilt" Houses on pages 12 and 13.

*For Our Easy Payment Plan See Page 144*

**The Rodessa Is Shown in Colors on Page 21**

**Can be built on a lot 27 feet wide**

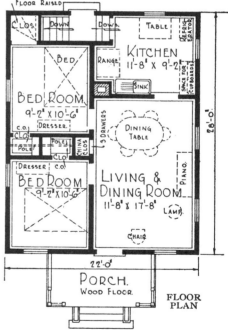

**No. P3203 Without Bathroom**

Designed for districts where no sewerage or cesspool facilities are available. Very much the same as No. P7041 but without bathroom.

### OPTIONS

*Sheet Plaster and Plaster Finish,* to take the place of wood lath, $111.00 extra for No. P7041 and $106.00 for No. P3203. See page 109.
*Oriental Asphalt Shingles,* guaranteed 17 years, instead of wood shingles, $22.00 extra.
*Storm Doors and Windows,* $36.00 extra for No. P7041 and $41.00 for No. P3203.
*Screen Doors and Windows,* galvanized wire, $26.00 extra for either design.

For prices of Plumbing, Heating, Wiring, Electric Fixtures and Shades see pages 130 and 131.

THE DIGNITY and beauty of the Dutch colonial exterior is combined in The Amsterdam Home with modern interior arrangement and conveniences. Each room has been carefully planned with a view toward providing generous space for the usual furniture and light on two sides.

### FIRST FLOOR

**The Living Room.** As you enter the living room from the tapestry brick paved front porch you are met with the pleasing view shown on the opposite page—see living room —a colonial mantel and fireplace, a beautiful colonial stairway, French doors leading to the music or sun room and French doors entering the dining room. A closet for coats fitted with a mirror door is provided at the foot of the stairway. Size of living room, 21 feet 2 inches by 14 feet 8 inches. There are two front windows and two side windows.

**The Music or Sun Room.** It has six pairs

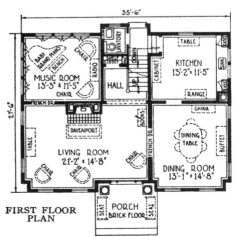

**FIRST FLOOR PLAN**

**Honor Bilt**

## The Amsterdam
### No. P13196A "Already Cut" and Fitted
## $3,641.00

of French windows. Size, 13 feet 3 inches by 11 feet 5 inches.

**The Dining Room.** Lighted by two front windows and two side windows. Size, 13 feet 1 inch by 14 feet 8 inches.

**The Kitchen** is equipped with our de luxe kitchen outfit, which includes built-in cupboards. Space for range and refrigerator.

The rear entry has a lavatory, stairs to basement and grade door.

### SECOND FLOOR

**The Bedrooms.** French windows light the stairs and hall, off which is provided a linen closet, the four bedrooms, and bathroom. Each of the bedrooms is provided with a clothes closet. The rear bedroom, left, has five pairs of French windows. It can be used for a sleeping porch and, if so, can accommodate three or four cots. Size, 13 feet 3 inches by 11 feet 3 inches. The other three bedrooms are lighted and aired by windows on two sides.

**The Bathroom** has a built-in medicine case and shelves for linen.

**The Basement.** Room for heating plant, laundry and furnace.

**Height of Ceilings.** First floor, 9 feet from floor to ceiling. Second floor, 8 feet 6 inches from floor to ceiling. Basement, 7 feet from floor to joists.

### What Our Price Includes

**At the price quoted we will furnish all the material to build this eight-room house, consisting of:**
**Lumber; Lath;**
**Roofing,** Best Grade Clear Red Cedar Shingles;
**Siding,** Clear Cypress or Clear Red Cedar, Bevel;
**Framing Lumber,** No. 1 Quality Douglas Fir or Pacific Coast Hemlock;
**Flooring,** Clear Maple for Kitchen, Lavatory and Entry, Clear Oak for Balance of First Floor. Tile for Bathroom, Clear Douglas Fir or Pacific Coast Hemlock for Balance of Second Floor;
**Finishing Lumber;**
**High Grade Millwork** (see pages 110 and 111);
**Interior Doors,** First Floor, French Doors, Mirror Door and One-Panel Birch and Douglas Fir Doors. Second Floor, Two-Cross Panel Design of Douglas Fir;
**Trim,** Birch for Living Room, Dining Room, Music Room and Stair Hall; Beautiful Grain Douglas Fir or Yellow Pine for All Other Rooms;

**Can be built on a lot 42 feet wide**
This house can be built with rooms reversed. See page 3.

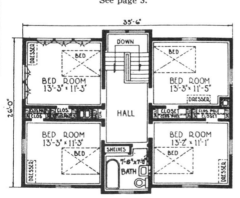

**SECOND FLOOR PLAN**

**Windows,** California Clear White Pine;
**Medicine Case;**
**Kitchen De Luxe Sink Outfit;**
**Mantel, Colonial; Colonial Shutters;**
**Eaves Trough and Down Spout;**
**40-Lb. Building Paper; Sash Weights;**
**Chicago Design Hardware** (see page 132);
**Paint for Three Coats Outside Trim and Siding;**
**Shellac, Varnish and Stain** for All Doors, Stair Treads and Stair Rail; Enamel for All Interior Trim and Woodwork on First Floor;
**Shellac and Varnish** for All Trim and Doors on Second Floor, except Bathroom; Enamel for Bathroom;
**Shellac, Wood Filler and Floor Varnish** for Oak and Maple Floors;
**Coal Chute.**
Complete Plans and Specifications.
We guarantee enough material to build this house. Price does not include cement, brick or plaster.
See description of "Honor Bilt" Houses on pages 12 and 13.

### OPTIONS

*Sheet Plaster and Plaster Finish, furnished in place of wood lath, $280.00 extra. See page 109.*

*Oriental Asphalt Shingles, guaranteed 17 years, instead of wood shingles, $60.00 extra.*

*Storm Doors and Windows, $108.00 extra.*

*Screen Doors and Windows, galvanized wire, $74.00 extra.*

For prices of Plumbing, Heating, Wiring, Electric Fixtures and Shades, see pages 130 and 131

**For Our Easy Payment Plan See Page 144**

# The Amsterdam Interiors

*A partial view of the Living Room, the Sun Room or Music Room and the Stairway*

**D**UTCH COLONIAL architecture is a much appreciated heritage from the dawn of our great Republic.

Its everlasting and growing popularity is due in no small measure to its beautiful design, practical layout, spacious rooms and economy of upkeep.

The Amsterdam is a fine example of ultra-modern Dutch colonial architecture. It is the kind of a home one readily learns to love.

"Honor Bilt" quality material and ready-cut construction make this home a better value and, besides, you save money on our low factory-to-you price.

*Coal Chute*

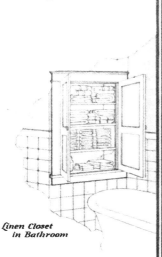

*All Bedrooms of The Amsterdam Have Good Spacing, Light and Air*

*Linen Closet in Bathroom*

*This Modern Kitchen is Planned to Save Time and Effort, Providing More Leisure for the Busy Housewife or Help. See Our Built-In Popular De-Luxe Outfit as Illustrated Above*

*A Rear view of the House showing the Trade Entry*

THE AVALON bungalow readily meets with favor wherever it is erected. Its architectural lines are artistic, the construction is very substantial and the material is of "Honor Bilt" quality.

Customers who have built the Avalon tell us freely that our approved ready-cut system has saved them much money. Yes, and many a handsome profit has been made by a customer when selling his Avalon.

The exterior is adorned with the fireplace chimney, gables, overhanging eaves and massive stucco porch pillars. The large L shape porch measures 18 feet 6 inches by 20 feet. It may be screened and furnished as suggested in the illustration below.

**The Living Room.** Pleasantly located in the front of the house are both the living and dining rooms. Each of these rooms has four pairs of French casement sash windows. You will like these windows because of their fine appearance. But it will take a hot summer's day to really appreciate how enjoyable it is to open all the window space instead of just half, as usual.

Size of living room, 14 feet 3 inches by 16 feet 8 inches. The fireplace and mantel is the center of attraction, as it is built in the middle of the front wall, with French windows on either side. There is plenty of space for furniture, including liberal wall space for a piano, davenport, etc. A coat closet for guests is in one corner.

**The Dining Room.** A book cased opening unites the living room and dining room as one large room. Here friends may be entertained, assuring unusually large space for social gatherings, considering the house is only 28 by 48 feet. Dining room dimensions are 12 feet 7 inches by 14 feet 2 inches, and easily accommodates a dining set with wall space for buffet.

**The Kitchen.** The dining room leads to a swinging door, which opens directly into the kitchen. Size of kitchen, 14 feet 3 inches by 10 feet 11 inches. The sink and the built-in cupboard are together, and are between the dining room door and the two windows. This means light and air, and hundreds of steps saved every day when preparing meals. The space for the range is on

**Front Porch When Screened**

## Honor Bilt

# The Avalon
### No. P13048 "Already Cut" and Fitted
## $2,530.00

the adjoining wall where the light will shine directly into the oven. Close by is space for a table and chair. A door opens into the rear entry, which has space for refrigerator and space for shelving; then, too, the rear entry enables the housewife to keep the kitchen clean and warmer. Here steps lead to basement and a glazed door to the yard.

**The Bedrooms.** Right off the dining room is a hallway that connects with the three bedrooms, bathroom, linen closet and stairway to attic. Each bedroom has two windows and a clothes closet. In the largest of the bedrooms there is space for twin beds, besides the dresser, chiffonier, night table and chairs.

**The Bathroom.** Has a built-in medicine cabinet, and plumbing may be roughed-in on one wall.

**Basement.** Room for furnace, laundry and storage.

**Height of Ceilings.** Main floor, 9 feet from floor to ceiling. Basement, 7 feet from floor to ceiling.

### What Our Price Includes

At the price quoted we will furnish all the material to build this six-room house, consisting of:

**Lumber; Lath;**
**Roofing,** Oriental Slate Surfaced Shingles, 17-Year Guarantee;
**Siding,** Clear Cypress or Clear Red Cedar, Bevel;
**Framing Lumber,** No. 1 Quality Douglas Fir or Pacific Coast Hemlock;
**Flooring,** Clear Oak and Maple;
**Porch Ceiling,** Clear Douglas Fir or Pacific Coast Hemlock;
**Finishing Lumber;**
**High Grade Millwork** (see pages 110 and 111);
**Interior Doors,** One-Panel Design of Douglas Fir;
**Trim,** Beautiful Grain Douglas Fir or Yellow Pine;
**Windows,** California Clear White Pine;
**Medicine Case;**
**Kitchen Cupboard; Bookcases;**
**Brick Mantel; Flower Boxes;**
**Eaves Trough and Down Spouts;**
**40-Lb. Building Paper; Sash Weights;**
**Chicago Design Hardware** (see page 132);
**Paint,** Three Coats for Outside Trim and Siding;
**Stain,** Two Brush Coats for Shingles on Gable Walls;
**Shellac and Varnish** for Interior Trim and Doors;
**Shellac, Paste Filler and Floor Varnish** for Oak and Maple Floors.

Complete Plans and Specifications.

We guarantee enough material to build this house. Price does not include cement, brick or plaster. See description of "Honor Bilt" Houses on pages 12 and 13.

**Can be built on a lot 45 feet wide**

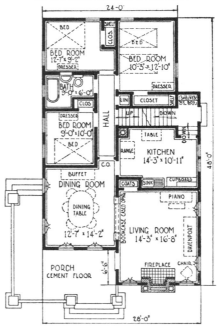

FLOOR PLAN

### OPTIONS

*Sheet Plaster and Plaster Finish to take the place of wood lath, $205.00 extra. See page 109.*

*Storm Doors and Windows, $79.00 extra.*

*Screen Doors and Windows, galvanized wire, $50.00 extra.*

*Oak Doors and Trim for living room and dining room, $94.00 extra.*

For Prices of Plumbing, Heating, Wiring, Electric Fixtures and Shades see pages 130 and 131.

### For Our Easy Payment Plan See Page 144

# FIVE ROOMS, BATH AND BIG PORCH

THE PRESCOTT colonial home has a restful and charming appeal. Much has been said and written about its simplicity, modern interior and unusually good construction.

**The Exterior** blends itself admirably to the use of wide colonial siding and terraced grade. Painted white or ivory, with green roof, it fully deserves the favorable comment one hears from neighbors and friends. Then, too, the front porch, size, 22 feet by 7 feet, is almost a living room during summer.

**The Vestibule** is entered from the front porch. A coat closet for guests is an appreciated convenience here.

**The Living Room.** A door divides the vestibule and the living room. At the left is located a large closet commonly used to conceal a door bed, giving this home the use of three bedrooms in an otherwise two-bedroom plan. Plenty of wall space for piano and furniture. An attractive set of three windows in the front, and a single high sash at the side, admit light and air. Floor space measures 16 feet 8 inches by 11 feet 5 inches.

**The Dining Room** and the living room connect through a wide cased opening. Size, 11 feet 5 inches by 10 feet 7 inches. A double window at the side reflects light and permits ventilation.

**The Kitchen.** A swinging door from the dining room connects with the kitchen. Size of kitchen, 7 feet 10 inches by 10 feet 5 inches. The right wall is devoted to space for sink with a cupboard on either side. Space for range and refrigerator is provided on the left wall. A window admits fresh air and light. At the rear end a door connects with rear entry, which has stairs to basement and grade.

**The Bedrooms.** One bedroom opens off the dining room. It is 11 feet 5 inches by 10 feet 7 inches. A double window insures light and ventilation. The rear bedroom opens off the hall from the dining room. It has two windows.

**The Bathroom,** too, opens off the hall. Plumbing fixtures can be roughed-in on one wall, saving on installation cost. It has a built-in medicine cabinet and a window.

**The Basement.** Excavated basement under the entire house. Room for storage, furnace and laundry.

**Height of Ceilings.** Main floor, 8 feet 6 inches from floor to ceiling. Ridge, 7 feet 6 inches from ceiling joists to ridge. Basement, 7 feet from floor to joists.

## What Our Price Includes

At the price quoted we will furnish all the material to build this five-room house consisting of:
**Lumber; Lath;**
**Roofing,** Best Grade Clear Red Cedar Shingles;
**Siding,** Clear Cypress or Clear Red Cedar, Bevel;
**Framing Lumber,** No. 1 Quality Douglas Fir or Pacific Coast Hemlock;
**Flooring,** Clear Grade Douglas Fir or Pacific Coast Hemlock;
**Porch Flooring,** Clear Grade Douglas Fir or Pacific Coast Hemlock;
**Porch Ceiling,** Clear Grade Douglas Fir or Pacific Coast Hemlock;
**Finishing Lumber;**
**High Grade Millwork** (see pages 110 and 111);
**Interior Doors,** Two Panel Design of Douglas Fir;
**Trim,** Beautiful Grain Douglas Fir or Yellow Pine;
**Windows** of California Clear White Pine;
**Medicine Case;**
**Kitchen De Luxe Outfit;**
**Eaves Trough and Down Spout;**
**40-Lb. Building Paper; Sash Weights;**
**Chicago Design Hardware** ( see page 132);
**Paint** for Three Coats Outside Trim and Siding;
**Varnish and Shellac** for Interior Doors and Trim.

We guarantee enough material to build this house. Price does not include cement, brick or plaster.

See description of "Honor Bilt" Houses on pages 12 and 13 of Modern Homes Catalog.

### OPTIONS

*Sheet Plaster and Plaster Finish, to take the place of wood lath, $141.00 extra. See page 109.*

*Storm Doors and Windows, $67.00 extra.*

*Screen Doors and Windows, galvanized wire, $48.00 extra.*

*Oriental Slate Surfaced Shingles, in place of wood shingles, for roof, $35.00 extra.*

*Attic Stairs and 1x6 Dressed and Matched Floor $120.00 extra.*

For prices of Plumbing, Heating, Wiring, Electric Fixtures and Shades see pages 130 and 131.

**Honor Bilt**

## The Prescott
No. P3240 "Already Cut" and Fitted
### $1,873.00

**Can be built on a lot 30 feet wide**

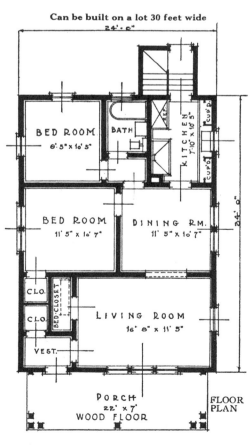

*For Our Easy Payment Plan See Page 144*

# SIX ROOMS, BATH AND PORCH

THE SALEM is a home combining character and loveliness. The top part of house and porch gable are paneled on a stucco background. The balance of the exterior walls are finished with bevel siding. The front porch is a gem in the comfort it gives during the warm months. It can be glazed or screened in, making it usable the year around. Entrance to the porch is gained only from the interior through the living room.

### FIRST FLOOR

**The Reception Hall.** Entering the hall from the side porch you face the staircase to the second floor. To the side of stairs is a coat closet with hat shelf.

**The Living Room.** To the left of hall a large cased opening admits you to the large living room that is 21 feet 3 inches long by 12 feet 6 inches wide. Perfect balance of features makes the living room beautiful. See floor plan. Balancing the pair of French doors to the front porch is the mantel and fireplace with its tile hearth.

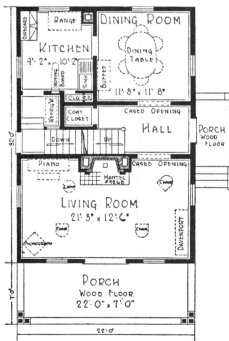

FIRST FLOOR PLAN

---

**Honor Bilt**

## The Salem
### No. P3211 "Already Cut" and Fitted
## $2,634.00

The living room occupies the entire width of the front part of house. Two windows on the front and one window on each side insure enough sunshine and air. Furniture and piano have ample space here. This room presents a very striking appearance.

**The Dining Room.** To the right of hall a wide cased opening shows the dining room. Here a dining set including buffet fits becomingly and will accommodate guests. Two windows admit light and air. Size of dining room, 11 feet 8 inches by 11 feet 8 inches.

**The Kitchen.** A swinging door connects the dining room and kitchen. Here the housewife or maid will find pleasure in preparing the daily meals and baking. The cupboards, the kitchen closet, clothes chute, the sink, and space for gas range are placed with a view of saving the housewife many steps. A built-in ironing board that opens out when needed and concealed by a panel when not in use is an added convenience.

Door opens into side entry where space for refrigerator is provided. This entry has stairs to basement and sash door to grade and outside.

### SECOND FLOOR

**The Bedrooms.** A beautiful stair hall, lighted by a window on stair landing, connects with the three bedrooms, bath and linen closet. A clothes chute leading to laundry is also conveniently located in the hall. The main bedroom fronts the entire front part of the house. Here is a bedroom that surely will appeal to those desiring a spacious and commodious chamber. Plenty of space for two full size beds, dresser, chiffonier, table and lamp and chairs. Two large closets, each with hat shelf and clothes pole are at your service. Two front windows and one window on each side provide light and ventilation. The two rear bedrooms each have good size closets with clothes poles and hat shelves. Windows on two sides give the bedrooms all the needed sunshine and air.

**The Basement.** Room for furnace, laundry and storage.

**Height of Ceilings.** First floor, 9 feet from floor to ceiling. Second floor, 8 feet from floor to ceiling. Basement, 7 feet from floor to joists.

### What Our Price Includes

At the price quoted we will furnish all the material to build this six-room two-story house, consisting of:
Lumber; Lath;
Roofing, Oriental Slate Surfaced Shingles, 17-Year Guarantee;
Siding, Clear Cypress or Clear Red Cedar, Bevel;
Framing Lumber, No. 1 Quality Douglas Fir or Pacific Coast Hemlock;
Flooring, Clear Douglas Fir or Pacific Coast Hemlock;
Porch Flooring, Clear Edge Grain Douglas Fir;
Porch Ceiling, Clear Douglas Fir or Pacific Coast Hemlock;
Finishing Lumber;
High Grade Millwork (see pages 110 and 111);
Interior Doors, Two Panel Design of Douglas Fir;

---

**Can be built on a lot 33 feet wide**

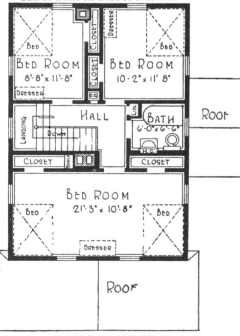

SECOND FLOOR PLAN

Trim, Beautiful Grain Douglas Fir or Yellow Pine;
Windows, California Clear White Pine;
Medicine Case;
Kitchen Cupboards;
Built-in Ironing Board;
Colonial Mantel;
Eaves Trough; Down Spout;
40-Lb. Building Paper; Sash Weights;
Stratford Design Hardware (see page 132);
Paint for Three Coats Outside Trim and Siding;
Shellac and Varnish for Interior Trim and Doors.
Complete Plans and Specifications.
We guarantee enough material to build this house. Price does not include cement, brick or plaster.
See description of "Honor Bilt" Houses on pages 12 and 13.

### OPTIONS

*Sheet Plaster and Plaster Finish, to take the place of wood lath and plaster, $206.00 extra. See page 109.*
*Storm Doors and Windows, $88.00 extra.*
*Screen Doors and Windows, galvanized wire, $58.00 extra.*

For prices of Plumbing, Heating, Wiring, Electric Fixtures and Shades see pages 130 and 131.

---

**For Our Easy Payment Plan See Page 144**

# The SALEM ~ INTERIORS

*UPPER LEFT*—With two large windows, The Salem kitchen is bright and attractive. Built-in features make it convenient and easy to keep clean.

*CENTER*—A view of the living room which extends the entire width of the house. The fireplace with its tile hearth makes the living room enjoyable and admired.

*LOWER LEFT*—Quiet and restful bedrooms are a feature of The Salem Home. The bedroom shown is large and airy. The spacious size of this room permits the ideal arrangement of bedroom furniture.

*UPPER RIGHT*—Hospitality is reflected in the well planned dining room. The spacing and lighting of The Salem dining room makes it appealing and inviting.

*LOWER RIGHT*—The reception hall suggests a home of refinement. A coat closet and hat shelf is conveniently located to the side of the stairs.

All arrangements shown are merely suggestive. With The Salem Home you will be able to express your own taste in home furnishings. Each room is so planned that it lends itself well to any number of fine decorative plans.

THE MARTHA WASHINGTON is a design that will delight lovers of the colonial type of architecture.

The entrance with its colonial columns, door and side lights is most inviting and attractive, flanked by the double colonial windows on either side. The pure white wide siding is set off to the best possible advantage by the sea green Oriental Shingles and colonial chimney. Fancy this commodious home set in its proper landscape! The red tile approach to the brick porch ornamented with flowers and shrubbery will often prompt you to stand and admire this beautiful creation of the architect's skill.

The view to the visitor or passerby presents a vision of hospitality and brightness that is characteristic of many famous historical homes. Study the floor plans. Compare the size of rooms with houses you are familiar with of much larger dimensions and you will appreciate that the unusual economical planning gives the Martha Washington the same size living rooms as found in houses of nearly double its size.

### FIRST FLOOR

**The Living Room.** Opening the colonial door you enter a spacious living room extending the entire width of the house. Floor dimensions are: 27 feet 8 inches by 14 feet 5 inches. Directly in front is a colonial brick mantel. To your left is a double bay window with hinged seat and bookcase. To the right an artistic open stairway to the second floor with a coat closet on the landing. In this big room you have ample space for a large piano, radio, davenport, library table, plenty of chairs, reading lamp and other furniture you deem necessary for your own comfort and the entertainment of your guests. This room is well lighted and ventilated.

**The Dining Room.** To the left of the brick mantel in the living room French doors lead into the dining room which is lighted by a pair of large windows; also two windows to the left, between which is a bay large enough to accommodate a commodious buffet. Note the good wall spaces for furniture. It measures 15 feet 3 inches by 12 feet 5 inches.

**The Kitchen.** A swinging door from the dining room leads to the modern kitchen—11 feet 10 inches by 12 feet 5 inches—which is the particular pride of every Martha Washington owner. Our Kitchen De Luxe outfit, disappearing ironing board, the stairway which enables one to reach the second floor from either living room or kitchen, the extra closet on kitchen stair landing, and the modern grade entrance with refrigerator space inside the house—but outside the kitchen—are unusual and most desirable features. Plenty of space for stove and table. Plenty of light and air.

### SECOND FLOOR

**The Bedrooms.** There are three bedrooms, bathroom and sleeping porch on this floor. The front and rear bedrooms at the left have commodious wardrobe closets. See pages 110 and 111. The bathroom is convenient to all rooms, and there is a handy linen closet off the hall. The lighting and ventilating arrangement is perfect. The sleeping porch has plenty of

room for two beds, usable during the winter as well as the summer. You will appreciate the added comfort and healthfulness of the sleeping porch during the hot summer nights.

**The Basement.** Excavated basement under the entire house, lighted with basement sash. Room for heating plant, laundry and storage.

**Height of Ceilings.** First floor, 9 feet from floor to ceiling. Second floor, 8 feet 2 inches from floor to ceiling. Basement, 7 feet from floor to joists.

### What Our Price Includes

At the price quoted we will furnish all the material to build this seven-room, two-story house, consisting of:
Lumber; Lath;
**Roofing,** Oriental Slate Surfaced Shingles, 17-Year Guarantee;
**Siding,** Clear Cypress or Clear Red Cedar, Drop;
**Framing Lumber,** No. 1 Quality Douglas Fir or Pacific Coast Hemlock;
**Flooring,** Clear Oak and Maple for First Floor, Clear Douglas Fir or Pacific Coast Hemlock for All of Second Floor, Except Bathroom; Tile Floor in Bathroom;
**Porch Ceiling,** Clear Douglas Fir or Pacific Coast Hemlock;
**Finishing Lumber;**
**High Grade Millwork** (see pages 110 and 111);
**Interior Doors,** One-Panel Design for First Floor and Two-Panel Design for Second Floor. All of Douglas Fir;
**Trim,** Beautiful Grain Douglas Fir or Yellow Pine;
**Windows,** California Clear White Pine;
**Medicine Case; Wardrobes;**
**Kitchen Cupboards;**
**Combination Seat and Bookcases;**
**Built-In Ironing Board;**
**De Luxe Tile Sink Outfit;**
**Mantel, Colonial;**
**Eaves Trough and Down Spout;**
**40-Lb. Building Paper;**
**Sash Weights;**
**Chicago Design Hardware** (see page 132);
**Paint for Three Coats Outside Trim and Siding;**
**White Enamel for Interior Trim;**
**Stain, Shellac and Varnish** for Interior Doors;
**Shellac, Paste Filler and Floor Varnish** for Oak and Maple Floors.
Complete Plans and Specifications.
We guarantee enough material to build this house. Price does not include cement brick or plaster. See description of "Honor Bilt" Houses on pages 12 and 13.

### OPTIONS

*Sheet Plaster and Plaster Finish, instead of wood lath, $268.00 extra. See page 109.*
*Storm Doors and Windows, $120.00 extra.*
*Lawn Seat, $15.00 extra.*
*Screen Doors and Windows, galvanized wire, $72.00 extra.*

---

## The Martha Washington
### No. P13080A "Already Cut" and Fitted
## $3,727.00

**Can be built on a lot 40 feet wide**

FIRST FLOOR PLAN

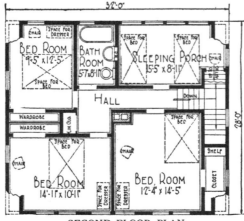

SECOND FLOOR PLAN

For Prices of Plumbing, Heating, Wiring, Electric Fixtures and Shades see pages 130 and 131.

---

*For Our Easy Payment Plan See Page 144*

## Honor Bilt

## The Elsmore
### No. P13192 "Already Cut" and Fitted
### $2,391⁰⁰

THE ELSMORE is the better grade of bungalow. For comfort, convenience and pleasing design, it can't be beat! Each time it is built its proud owner acclaims —"It's the best house on my street!" The Elsmore's beautiful exterior, plus a well arranged interior, insures satisfaction in the highest degree. Careful planning and good material at direct-from-factory prices make this attractive bungalow an unusually good bargain.

**The Exterior.** Observe the combination hip and gable roof. The overhang of gable has huge bracket supports. Panel adornments over stucco in the gable, triple bungalow columns, decorative porch rail and twin piers, are other features which are much appreciated. The fireplace chimney on the outside adds to the charm. The porch, 25 feet by 8 feet, is ample size for porch swing and furniture. When screened or glazed it is serviceable the year around.

**Can be built on a lot 46 feet wide**

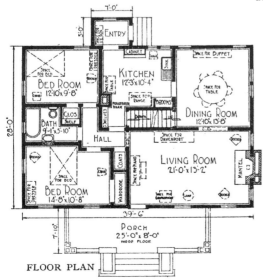

**FLOOR PLAN**

**The Living Room.** Size, 21 feet by 13 feet 2 inches. Here liberal space offers a charming setting for the cozy mantel and fireplace, which is provided with a high sash on both sides. Wall space for a piano and living room furniture makes possible any number of pleasing arrangements. Two front windows admit ample light and air.

**The Dining Room.** From the living room a cased opening connects with the dining room. Size, 12 feet 10 inches by 13 feet 8 inches. There is wall space between the two rear windows for a buffet, and room for table and chairs, leaving plenty of space for serving. Here cheerfulness abounds. A double window at the side provides light and cross ventilation. Stairs to the attic are reached through door from dining room.

**The Kitchen.** A swinging door opens from the dining room into the modern kitchen. Size, 12 feet 3 inches by 10 feet 4 inches. The location of sink, range and kitchen table is planned to meet the latest approved ideas. It has a convenient kitchen cabinet (see pages 110 and 111). A broom closet with two doors is in a handy corner. Light and ventilation from a double window. From the kitchen a stairway leads to the basement. A door opens from the kitchen into the rear entry, where the icebox is to be kept. Stairs lead to the grade door.

**The Bedrooms.** From the living room a hallway with a clothes closet connects with the two bedrooms and bath. The front bedroom. size 14 feet 8 inches by 10 feet 8 inches, has a commodious wardrobe with two doors. Two front windows and one side window provide plenty of light and air. The rear bedroom, size 12 feet 10 inches by 9 feet 8 inches, has a clothes closet. There is a window on each outside wall.

The bathroom has a built-in medicine case, and is lighted and ventilated by a window.

**Basement.** Room for a furnace, laundry and storage.

**Height of Ceilings.** Main floor, 9 feet from floor to ceiling. Basement, 7 feet from floor to joists. Concrete floor.

### What Our Price Includes

At the price quoted we will furnish all the material to build this five-room bungalow, consisting of:

**Lumber; Lath;**

**Roofing,** Best Grade Clear Red Cedar Shingles;

**Siding,** Clear Cypress or Clear Red Cedar, Bevel;

**Framing Lumber,** No. 1 Quality Douglas Fir or Pacific Coast Hemlock;

**Flooring,** Clear Maple for Kitchen and Bathroom, Clear Oak for Other Rooms;

**Porch Flooring,** Clear Edge Grain Fir;

**Porch Ceiling,** Clear Douglas Fir or Pacific Coast Hemlock;

**Finishing Lumber;**

**High Grade Millwork** (see pages 110 and 111);

**Interior Doors,** Two Vertical Panel Design of Douglas Fir;

**Trim,** Beautiful Grain Douglas Fir or Yellow Pine;

**Windows,** California Clear White Pine;

**Medicine Case; Kitchen Cabinets;**

**Mantel;**

**Eaves Trough; Down Spouts;**

**40-Lb. Building Paper; Sash Weights;**

**Chicago Design Hardware** (see page 132);

**Paint** for Three Coats Outside Trim and Siding;

**Shellac and Varnish** for Interior Trim and Doors;

**Shellac, Paste Filler and Floor Varnish** for Oak and Maple Floors.

Complete Plans and Specifications.

We guarantee enough material to build this house. Price does not include cement, brick or plaster. See description of "Honor Bilt" Houses on pages 12 and 13.

### OPTIONS

*Sheet Plaster and Plaster Finish, to take the place of wood lath, $179.00 extra. See page 109.*

*Oriental Asphalt Shingles, instead of wood shingles, $51.00 extra.*

*Oak Doors and Trim in living room and dining room, $87.00 extra.*

*Storm Doors and Windows, $69.00 extra.*

*Screen Doors and Windows, galvanized wire, $43.00 extra.*

For prices of Plumbing, Heating, Wiring, Electric Fixtures and Shades see pages 130 and 131.

*For Our Easy Payment Plan See Page 144*

**Honor Bilt**

# The Hamilton
### No. P3200 "Already Cut" and Fitted
## $2,124.00

THE HAMILTON bungalow fulfills all the promises of its handsome exterior. It is just as up to date and has as many good features in the interior. Just a glance at the floor plan will reveal the excellent arrangement of the rooms. Then, again, the interior views on the opposite page show how tastefully one can furnish this home. Truly the Hamilton is a model for appearance, convenience and price. Of course, the material and construction are "Honor Bilt."

**The Exterior** is adorned with a dormer, fireplace chimney and thick cedar shingle roof. The porch is included under the main roof, and is easily changed to a sun room by screening and glazing. Size of porch, 10 feet 6 inches by 8 feet.

**The Living Room.** Size, 12 feet 9 inches by 18 feet 5 inches. A mantel and fireplace is set in the center of the left outer wall. There is plenty of space for furniture and wall space for a piano. The effect of a sun parlor is obtained at a small expense and without the necessity of additional furniture. Six French windows flood this room with sunshine and air.

**The Dining Room** is connected with the living room by a wide cased opening. Both rooms are practically one large room, thus making entertaining very simple. Size of dining room, 14 feet 3 inches by 10 feet 10 inches. It has two windows in a recessed bay with two hinged seats.

**The Kitchen.** Size, 12 feet 9 inches by 10 feet 1 inch. It is equipped with our beautiful Kitchen De Luxe outfit. The Priscilla home, see page 34, and many of our best homes are also equipped with our Kitchen De Luxe outfit.

The Hamilton kitchen has additional room for a range, table, etc. It connects with the dining room through a swinging door. There is a pantry through another door, a broom closet, and a rear entry with space for a refrigerator; also stairs to basement and grade.

**The Breakfast Alcove.** A cased opening leads from the kitchen to the alcove. It has a built-in table and two benches, and a window.

**The Bedrooms.** The front bedroom opens off the living room. The rear bedroom and the bathroom open off the hallway from the dining room. Each bedroom has a clothes closet and two windows.

The bathroom has a built-in medicine case and a window.

**The Basement.** Room for furnace, laundry and storage.

**Height of Ceilings.** Main floor, 9 feet from floor to ceiling. Basement, 7 feet from floor to joists.

### What Our Price Includes

**At the price quoted we will furnish all the material to build this five-room bungalow, consisting of:**

**Lumber; Lath;**

**Roofing,** Best Grade Red Cedar Shingles;

**Siding,** Clear Cypress or Clear Red Cedar, Bevel;

**Framing Lumber,** No. 1 Quality Douglas Fir or Pacific Coast Hemlock;

**Flooring,** Clear Douglas Fir or Pacific Coast Hemlock;

**Porch Flooring,** Clear Edge Grain Fir;

**Porch Ceiling,** Clear Douglas Fir or Pacific Coast Hemlock;

**Finishing Lumber;**

**High Grade Millwork** (see pages 110 and 111);

**Interior Doors,** Inverted Two-Panel Design of Douglas Fir;

**Trim,** Beautiful Grain Douglas Fir or Yellow Pine;

**Windows,** California Clear White Pine;

**Medicine Case;**

**Kitchen De Luxe Sink Outfit;**

**Brick Mantel;**

**Breakfast Alcove and Seats;**

**Eaves Trough; Down Spouts;**

**40-Lb. Building Paper; Sash Weights;**

**Stratford Design Hardware** (see page 132);

**Paint** for Three Coats Outside Trim and Siding;

**Shellac and Varnish** for Interior Trim and Doors.

Complete Plans and Specifications.

We guarantee enough material to build this bungalow. Price does not include cement, brick or plaster. See description of "Honor Bilt" Houses on pages 12 and 13.

For prices of Plumbing, Heating, Wiring, Electric Fixtures and Shades see pages 130 and 131.

**Can be built on a lot 32 ft. wide**

FLOOR PLAN

#### OPTIONS

*Sheet Plaster and Plaster Finish to take the place of wood lath, $165.00 extra. See page 109.*

*Oriental Asphalt Shingles, guaranteed 17 years, instead of wood shingles, $34.00 extra.*

*Oak Doors, Trim and Floors in living and dining room, Maple Floors in kitchen and bathroom, $173.00 extra.*

*Storm Doors and Windows, $71.00 extra.*

*Screen Doors and Windows, galvanized wire, $45.00 extra.*

*For Our Easy Payment Plan See Page 144*

*Above*—This dining room breathes an air of hospitality, so dear to the heart of every home lover.

Harmonious furnishings from Sears, Roebuck and Co. only serve to verify the beauty of this room.

*Below*—A spotless kitchen! There is a place for every kitchen need in these built-in cases and bins. Saves hundreds of steps in the daily preparation of meals.

*The Living Room*—Sunlight comes in the window during the day and in the evening the firelight flickers on the hearth. There is an air of fellowship that makes this a real home.

"Home is a place of peace and devotion, simplicity and comfort, charity and protection."

—JEANNETTE SHERIDAN.

*Above*—The bedroom is comfortable, airy and flooded with sunshine.

*Below*—The colorful built-in breakfast room appeals to the family. Eating here of a morning adds zest to the meal.

## The Amhurst

### No. P3244 "Already Cut" and Fitted

## $2,825.00

THE AMHURST, a two-story quaint Dutch colonial home, is finished in brick veneer on the first floor and in wide siding on the second floor. It is greatly admired for, (1) picturesque and distinctive beauty, (2) Dutch colonial lines with modern interior, (3) substantial strength and permanence.

Besides the combination of brick veneer, the exterior is made even more pleasing by the dark green shutters, the red fireplace chimney, sloping roof lines and the inset front porch with trellis railings.

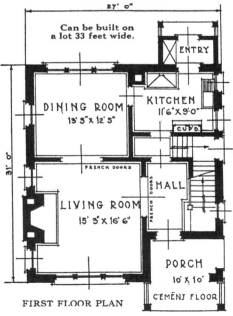

**Can be built on a lot 33 feet wide.**

27' 0"

ENTRY

DINING ROOM
13' 5" x 12' 5"

KITCHEN
11' 6" x 9' 0"

CUPD.

31' 0"

FRENCH DOORS

HALL

LIVING ROOM
15' 3" x 16' 6"

PORCH
10' x 10'

CEMENT FLOOR

FIRST FLOOR PLAN

The arrangement of the interior is a delight to its owner. Every room has been planned with careful thought to the other rooms, resulting in convenience, maximum livable space and cheerfulness.

### FIRST FLOOR

**The Reception Hall.** You enter the reception hall from the inset porch. Here is a semi-open stairway to the second floor, a door to side entry that connects with the kitchen, and French doors that open into the living room.

**The Living Room.** Floor dimensions: 15 feet 3 inches by 16 feet 6 inches. Directly opposite the French doors from the hall is a colonial mantel and fireplace, on either side of which is a high sash with space underneath for bookcases or other articles. A triple front window assures plentiful light and air. The wall and floor area accommodates a piano and a well furnished room.

**The Dining Room.** French doors connect the living room and the dining room. When these doors are open both rooms can be used as one room when entertaining. The dining room measures 13 feet 5 inches by 12 feet 5 inches. It has one side window and a triple window on the rear wall.

**The Kitchen** is 11 feet 6 inches by 9 feet. Very practically arranged so as to give greatest ease and working efficiency, everything seems so handy! It has a built-in cupboard, conveniently located near the sink. On the rear wall is space for the range. Two windows admit light and air. A door connects with the rear entry, which has space for a refrigerator, and for storage of brooms, etc. Another door connects with the side entry which has stairs to basement and grade.

### SECOND FLOOR

**The Bedrooms.** The stairway from the first floor leads to the short hall on the second floor. This hall connects with the three good size bedrooms and bathroom. Each bedroom has a clothes closet, and two windows, providing cross ventilation and plenty of light. Good attic space is provided with stairs leading from right front bedroom.

**The Bathroom** is exceptionally large for this size home. It has a built-in medicine cabinet, and side window, and a door that opens to deck over the rear entry.

**The Basement.** Room for furnace, laundry and storage.

**Height of Ceilings:** First floor, 8 feet 6 inches from floor to ceiling. Second floor, 8 feet 6 inches from floor to ceiling. Basement: 7 feet from floor to joists.

### What Our Price Includes

At the price quoted, we will furnish all the material to build this six-room house, consisting of:

**Lumber; Lath;**

**Roofing,** Best Grade Clear Red Cedar Shingles;

**Siding,** Clear Cypress or Clear Red Cedar, Bevel;

**Framing Lumber,** No. 1 Quality Douglas Fir or Pacific Coast Hemlock;

**Flooring,** Clear Grade Douglas Fir or Pacific Coast Hemlock;

**Porch Ceiling,** Clear Grade Douglas Fir or Pacific Coast Hemlock;

**Finishing Lumber;**

**High Grade Millwork** (see pages 110 and 111);
**Interior Doors,** Two-Panel Design of Douglas Fir;
**Trim,** Beautiful Grain Douglas Fir or Yellow Pine;
**Windows** of California Clear White Pine;
**Medicine Case; Kitchen Cabinets;**
**Mantel; Colonial Shutters;**
**Eaves Trough and Down Spout;**
**40-Lb. Building Paper; Sash Weights;**
**Chicago Design Hardware** (see page 132);
**Paint for Three Coats Outside Trim and Siding;**
**Varnish and Shellac for Interior Doors and Trim.**

We guarantee enough material to build this house. Price does not include cement, brick or plaster.

See description of "Honor Bilt" Houses on pages 12 and 13 of Modern Homes Catalog.

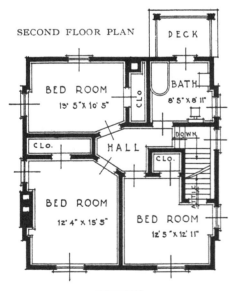

SECOND FLOOR PLAN

DECK

BED ROOM
13' 5" x 10' 5"

BATH
8' 5" x 8' 11"

CLO.

CLO.

HALL

CLO.

DOWN

BED ROOM
12' 4" x 15' 5"

BED ROOM
12' 5" x 12' 11"

### OPTIONS

*Sheet Plaster and Plaster Finish,* to take the place of wood lath, $228.00 extra. See page 109.

*Storm Doors and Windows,* $113.00 extra.

*Screen Doors and Windows* galvanized wire, $82.00 extra.

*Oriental Slate Surfaced Shingles,* in place of wood shingles for roof, $36.00 extra.

For prices of Plumbing, Heating, Wiring, Electric Fixtures and Shades see pages 130 and 131.

### For Our Easy Payment Plan See Page 144

### The Van Page

No. P3242 "Already Cut" and Fitted

**$2,650.00**

THE VAN PAGE, colonial two-story home, is another splendid example of modern adaptation of early Dutch architecture. All of the important features that are of Dutch-American ancestry are combined with modern ideas. For instance: The Van Page includes a variation from the regular hooded entrance, commonly associated with architecture of this kind. A modern front porch, cleverly adapted into the ensemble, takes the place of the decorative hood. Additional points of interest to the visitor are the gambrel roof and red brick fireplace chimney, the dark green shutters and flower boxes, white or ivory body, below which is the colorful red brick foundation. The entrance includes beautiful colonial sidelights.

You will be impressed with the convenient and compact arrangement of the interior. Every inch of space is used to the best advantage. Every room is spaced to allow the most usable room, in direct relation with the needs of a happy family.

"Honor Bilt" throughout, assuring quality and best value.

**Can be built on a lot 30 feet wide**

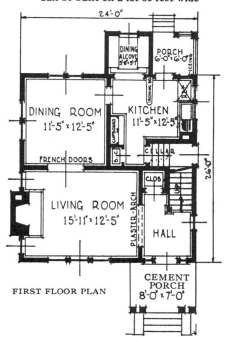

FIRST FLOOR PLAN

#### FIRST FLOOR

**Reception Hall** is separated from the living room by a cased opening. At the right a semi-open stairway leads to the second floor. A coat closet is located at the farther end of the hall. The hall is lighted by a high sash.

**The Living Room** measures 15 feet 11 inches by 12 feet 5 inches. A colonial fireplace and mantel is in the left wall. There is wall space for a piano and furniture. Three front windows and two side windows flood this comfortable living room with sunshine.

**The Dining Room** connects with the living room through French doors. Both rooms can be used as one large room when the occasion demands. Size, 11 feet 5 inches by 12 feet 5 inches. It has a double window at the side and a double window at the rear.

**The Kitchen** is entered from the dining room through a swinging door. Size, 11 feet 5 inches by 12 feet 5 inches. The housewife will be pleased with the arrangement of spaces for the range, sink and refrigerator, which can be iced from the provision porch. Directly over space for refrigerator is a built-in cupboard. Then, there is a built-in disappearing ironing board, a built-in cupboard and a broom closet.

A door leads to side entry, which has steps to basement and grade. Another door leads to rear porch.

**Dining Alcove** is right off the kitchen through a plastered arch opening.

#### SECOND FLOOR

**The Bedrooms.** Stairs from the reception hall lead to the hall on the second floor. Here the hall connects the three bedrooms, bathroom, linen closet and door to deck. Each of the bedrooms has two windows, and a clothes closet. The bathroom has a built-in medicine case.

Attic space is provided with stairs leading from right front bedroom.

**The Basement.** Room for furnace, laundry and storage.

**Height of Ceilings.** First floor, 8 feet 6 inches from floor to ceiling. Second floor, 8 feet 6 inches from floor to ceiling. Basement, 7 feet from floor to joists.

#### What Our Price Includes

At the price quoted we will furnish all the material to build this six-room house, consisting of:

Lumber; Lath;
Roofing, Best Grade Clear Red Cedar Shingles;
Siding, Clear Cypress or Clear Red Cedar, Bevel;
Framing Lumber, No. 1 Quality Douglas Fir or Pacific Coast Hemlock;
Flooring, Clear Grade Douglas Fir or Pacific Coast Hemlock;
Rear Porch Flooring, Clear Grade Douglas Fir or Pacific Coast Hemlock;
Front Porch, Cement;
Porch Ceiling, Clear Grade Douglas Fir or Pacific Coast Hemlock;
Finishing Lumber;
High Grade Millwork (see pages 110 and 111);
Interior Doors, Two-Panel Design of Douglas Fir;
Trim, Beautiful Grain Douglas Fir or Yellow Pine;
Windows of California Clear White Pine;
Medicine Case; Kitchen Cabinets;
Mantel; Built-In Ironing Board;
Breakfast Alcove, Table and Seats; Colonial Shutters;
Eaves Trough and Down Spout;
40-Lb. Building Paper; Sash Weights;
Chicago Design Hardware (see page 132);
Paint for Three Coats Outside Trim and Siding;
Varnish and Shellac for Interior Doors and Trim.

SECOND FLOOR PLAN

We guarantee enough material to build this house. Price does not include cement, brick or plaster.

Description of "Honor Bilt" Houses on pages 12 and 13 of Modern Homes Catalog.

#### OPTIONS

*Sheet Plaster and Plaster Finish, to take the place of wood lath, $205.00 extra. See page 109.*

*Storm Doors and Windows, $130.00 extra.*

*Screen Doors and Windows, galvanized wire, $94.00 extra.*

*Oriental Slate Surfaced Shingles, in place of wood shingles, per roof, $32.00 extra.*

For prices of Plumbing, Heating, Wiring, Electric Fixtures and Shades see pages 130 and 131.

*For Our Easy Payment Plan See Page 144*

Can be built on a lot 30 feet wide

THE FULLERTON two-story home is popular because it meets the needs of so many people. Its advantages over similar types of residences are many, some of which are: (1) You get so much space for such a small outlay of money; (2) Adapts itself equally well to city lots or country estates; (3) "Honor Bilt" material and construction.

### FIRST FLOOR

**The Porch.** 24 feet by 8 feet. When screened or glazed it offers every comfort and privacy of a large sun room.

**The Living Room** is 19 feet 10 inches by 12 feet 2 inches. A colonial mantel and fireplace is opposite the entrance. The open stairway leading to the second floor is located immediately to the right of entrance. It has a coat closet. Think of having plenty of space for the piano, davenport, radio, table and chairs! Three front windows, two windows on the left side and a high sash directly above the stairway provide an abundance of sunshine and air.

**The Dining Room.** A wide cased opening connects the living room and the dining room. Size of the dining room, 11 feet 11 inches by 12 feet 8 inches. Accommodates a complete dining set. There are two

**Honor Bilt**

## The Fullerton
### No. P3205 "Already Cut" and Fitted
## $2,294.00

side windows and a high sash directly over the buffet space.

**The Kitchen.** Entered from either the dining room or the living room, as well as from the outside by a grade entrance which leads also to the basement.

The kitchen has built-in cabinets, and space for refrigerator, range, table, chairs. One window on each outer wall provides light and cross ventilation. Size of kitchen, 10 feet 11 inches by 12 feet 8 inches.

### SECOND FLOOR

**The Bedrooms.** The stairway leads to the second floor hall which connects with the three bedrooms, bathroom and linen closet. Each bedroom has a clothes closet. The corner bedroom, as the floor plan shows, has three windows; the other two bedrooms have two windows each.

The bathroom has a built-in medicine case and a window.

**Basement.** Room for furnace, laundry and storage.

**Height of Ceilings.** First floor, 9 feet from floor to ceiling. Second floor, 8 feet 2 inches from floor to ceiling. Basement, 7 feet from floor to ceiling.

### What Our Price Includes

At the price quoted we will furnish all the material to build this six-room two-story home, consisting of:

**Lumber; Lath;**

**Roofing,** Oriental Slate Surfaced Shingles, 17-Year Guarantee;

**Siding,** Clear Cypress or Clear Red Cedar, Bevel;

**Framing Lumber,** No. 1 Quality Douglas Fir or Pacific Coast Hemlock;

**Flooring,** Clear Douglas Fir or Pacific Coast Hemlock;

**Porch Flooring,** Clear Edge Grain Fir;

**Porch Ceiling,** Clear Douglas Fir or Pacific Coast Hemlock;

**Finishing Lumber;**

**High Grade Millwork** (see pages 110 and 111);

**Interior Doors,** Vertical Two-Panel Design of Douglas Fir;

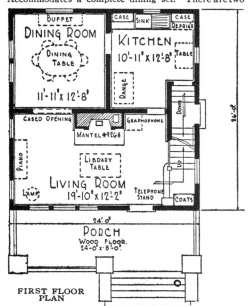

FIRST FLOOR PLAN

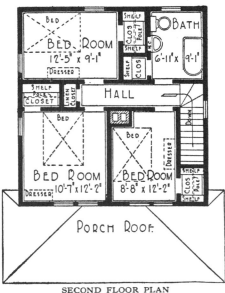

SECOND FLOOR PLAN

**Trim,** Beautiful Grain Douglas Fir or Yellow Pine;
**Windows,** California Clear White Pine;
**Medicine Case; Wardrobes;**
**Kitchen Cupboards;**
**Colonial Mantel;**
**Eaves Trough; Down Spout;**
**40-Lb. Building Paper; Sash Weights;**
**Stratford Design Hardware** (see page 132);
**Paint** for Three Coats Outside Trim and Siding;
**Stain** for Two Brush Coats for Shingles on Dormer Walls;
**Shellac and Varnish** for Interior Trim and Doors.

Complete Plans and Specifications.

We guarantee enough material to build this house. Price does not include cement, brick or plaster. See description of "Honor Bilt" Houses on pages 12 and 13.

### OPTIONS

*Sheet Plaster and White Plaster Finish,* to take the place of wood lath, $198.00 extra. See page 109.

*Storm Doors and Windows,* $76.00 extra.

*Screen Doors and Windows,* galvanized wire, $50.00 extra.

*Oak Stairs, Trim and Floor,* for living room and dining room, $191.00 extra.

For Prices of Plumbing, Heating, Wiring, Electric Fixtures and Shades see pages 130 and 131.

*For Our Easy Payment Plan See Page 144*

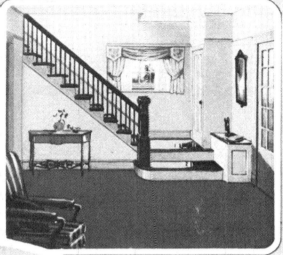

# The FULLERTON INTERIORS

*Above*—The architectural features of The Fullerton permit the use of any style of dining room furniture. The illustration shows how beautifully one style of furniture harmonizes with the dining room features.

*Below*—This ultra-modern kitchen is a treasure house of convenience to the housewife. See pages 110 and 111 for detailed description of the built-in cabinets.

*Center*—A spacious, well appointed living room is a source of constant pleasure to its owner. Here social gatherings enjoy a greater measure of hospitality, comfort and convenience.

"The threshold of happiness and success is at the door of every man's home."
—HERBERT FRANZ

*Above*—Another view of the living room, showing the open stairway to the second floor, and the coat closet on the landing.

*Below*—A colorful bedroom of pleasing qualities.

## The Albany

**No. P13199 "Already Cut" and Fitted**
### $2,232.00

**Can be built on a lot 23 feet wide**

**FIRST FLOOR PLAN**

WHILE THE ALBANY, being only 18 feet wide, is well suited for a narrow lot, it will look very well on a wider lot. The design is suitable for city, suburb or country. The interior arrangement is splendid. Its rooms are large, including living room, dining room, and front bedroom. The front porch, as shown in the picture, combined with siding and shingles, adds much to the appearance of this house.

### FIRST FLOOR

**The Porch.** Cement steps with brick buttress or railing lead to the spacious front porch. It measures 18 feet by 8 feet, and is almost a living room in the summer.

**The Living Room** opens from the porch. In the corner of the large living room there is a semi-open staircase to the second floor. Ample space provides for furniture. Light and air is provided from the sides and front. Size of living room, 17 feet 3 inches by 12 feet 5 inches.

**The Dining Room.** From the living room there is also a large cased opening into the dining room. Size, 13 feet 8 inches by 11 feet 7 inches. Plenty of room for a dining set. It has a guest clothes closet and a double window.

**The Kitchen.** Entry into the kitchen from the dining room is through a swinging door. In this kitchen will be found space to install every convenience for the housewife. There is space for sink, range and table. Two windows light and ventilate the kitchen. Door to the left opens to rear porch which has steps to yard. The side door (to right) connects with outside entrance which has stairs to basement.

### SECOND FLOOR

**The Bedrooms.** A hallway connects the three bedrooms and bath. The main room occupies the entire front part of the house and will accommodate twin beds if desired. It has three front windows and one side window. In the corner is a clothes closet with hat shelf. The center bedroom is lighted and aired from a double window. A clothes closet with hat shelf is close to space for the dresser. The rear bedroom has a clothes closet with hat shelf, and also a window that supplies light and air. The bathroom plumbing fixtures are on one wall.

**The Basement.** Under entire house. Room for furnace, laundry and storage.

**Height of Ceilings.** First floor, 9 feet from floor to ceiling. Second floor, 8 feet 6 inches from floor to ceiling. Basement, 7 feet from floor to joists.

### What Our Price Includes

At the price quoted we will furnish all the material to build this six-room two story house, consisting of:

**Lumber; Lath;**
**Roofing,** Oriental Slate Surfaced Shingles, 17-Year Guarantee;
**Siding,** Clear Cypress or Clear Red Cedar, Bevel;
**Framing Lumber,** No. 1 Quality Douglas Fir or Pacific Coast Hemlock;
**Flooring,** Clear Douglas Fir or Pacific Coast Hemlock for Interior Floors, Clear Edged Grain for Front and Rear Porch;
**Porch Ceiling,** Yellow Pine for Rear Porch;
**Finishing Lumber;**
**High Grade Millwork** (see pages 110 and 111);
**Interior Doors,** Two-Panel Design of Clear Douglas-Fir;

**Trim,** Beautiful Grain Douglas Fir or Yellow Pine;
**Windows,** California Clear White Pine;
**Medicine Case;**
**Eaves Trough and Down Spout;**
**40-Lb. Building Paper; Sash Weights;**
**Stratford Design Hardware** (see page 132);
**Paint for Three Coats Outside Trim and Siding;**
**Stain for Shingles on Dormer Walls and Gables for Two Brush Coats;**
**Shellac and Varnish** for Interior Doors and Trim.
Complete Plans and Specifications.

We guarantee enough material to build this house. Price does not include cement, brick or plaster. See description of "Honor Bilt" Houses on pages 12 and 13.

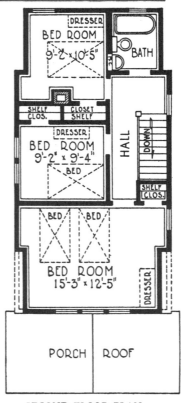

**SECOND FLOOR PLAN**

### OPTIONS

*Sheet Plaster and Plaster Finish, to take the place of wood lath, $189.00 extra. See page 109.*

*Storm Doors and Windows, $71.00 extra.*

*Screen Doors and Windows, galvanized wire, $52.00 extra.*

For prices of Plumbing, Heating, Wiring, Electric Fixtures and Shades see pages 130 and 131.

### For Our Easy Payment Plan See Page 144

THE NEW DUNDEE is an attractive bungalow at a very low price. It has an ornamental gabled porch with an inviting entrance. A trellis on the side to support vines. Our best quality material is used. In every way an excellent value.

**The Living Room.** Size, 12 feet 11 inches by 13 feet 5 inches. It has plenty of wall space for piano and furniture. Windows on two sides admit plenty of light and air.

**The Kitchen.** A door connects the living room and the large kitchen. Size of kitchen, 12 feet 11 inches by 9 feet 5 inches. Grouped together are the kitchen cabinet and space for gas range and sink. A window is directly opposite the stove to furnish light for oven. Space in the corner between the two windows will accommodate a dining table and chairs. Meals can be eaten here with perfect comfort.

A door from the kitchen connects with rear entry, which has space for ice box; steps to basement, and sash door to grade and outside.

**The Bedrooms.** A cased opening from the living room into a hall leads to the two bedrooms, bath and hall coat closet. Size of front bedroom, 9 feet 11 inches by 9 feet 11 inches. Size of rear bedroom, 9 feet 11 inches by 12 feet 5 inches.

Each bedroom is lighted and aired by two windows, and has a roomy clothes closet with hat shelf.

In the bathroom the plumbing may be placed on one wall, reducing installation costs. It has a medicine case.

**The Basement.** Space for furnace, laundry and storage.

**Height of Ceilings.** Main floor, 9 feet from floor to ceiling. Basement, 7 feet from floor to joists.

## Honor Bilt

### The Dundee
#### No. P3209 "Already Cut" and Fitted
### $1,405.00

### What Our Price Includes

At the price quoted we will furnish all the material to build this four-room bungalow, consisting of:

**Lumber; Lath;**
**Roofing,** Best Grade Clear Red Cedar Shingles;
**Siding,** Clear Cypress or Clear Red Cedar, Bevel;
**Framing Lumber,** No. 1 Quality Douglas Fir or Pacific Coast Hemlock;
**Flooring,** Clear Douglas Fir or Pacific Coast Hemlock;
**Porch Flooring,** Clear Edge Grain Fir;
**Porch Ceiling,** Clear Douglas Fir or Pacific Coast Hemlock;
**Finishing Lumber;**
**High Grade Millwork** (see pages 110 and 111);
**Interior Doors,** Five Cross Panel Design of Douglas Fir;
**Trim,** Beautiful Grain Douglas Fir or Yellow Pine;
**Windows** of California Clear White Pine;
**Medicine Case;**
**Kitchen Cupboard;**
**Eaves Trough and Down Spouts;**
**40-Lb. Building Paper; Sash Weights;**
**Stratford Design Hardware** (see page 132);
**Paint** for Three Coats Outside Trim and Siding;
**Shellac and Varnish** for Interior Doors and Trim.
Complete Plans and Specifications.

We guarantee enough material to build this bungalow. Price does not include cement, brick or plaster. See description of "Honor Bilt" Houses on pages 12 and 13.

Can be built on a lot 30 feet wide

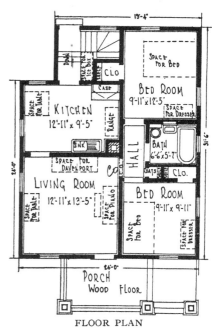

FLOOR PLAN

### OPTIONS

*Sheet Plaster and Plaster, Finish to take the place of wood lath, $122.00 extra. See page 109.*

*Oriental Asphalt Shingles, guaranteed 17 years, instead of wood shingles, $31.00 extra.*

*Storm Doors and Windows, $45.00 extra.*

*Screen Doors and Windows, galvanized wire, $29.00 extra.*

For prices of Plumbing, Heating, Wiring, Electric Fixtures and Shades see pages 130 and 131.

*For Our Easy Payment Plan See Page 144*

*Built by a Customer in Mt. Vernon on the Lake, N. Y.*

THE VERONA, a high class home of Dutch type of colonial architecture. Here's an "Honor Bilt" Home that always satisfies the owner and is adored by everyone in his locality. Its growing popularity is demonstrated by the ease with which it is sold after it is completed. Its simplicity and beauty make it a classic in architecture.

If you love Dutch colonial architecture, remember that it has been generally conceded by authorities that the Verona ranks first for price, quality and design.

Built many times in exclusive suburbs of New York, Chicago, Washington, Cleveland, Pittsburgh, Cincinnati and other large cities.

Aside from its rare beauty, it represents one of the best values in this catalog. At its present low price, it is a decided bargain!

Study its floor plans shown on opposite page. Could you desire more spacious rooms? Every inch is put to practical use. But don't overlook the many 1926 features; they are seldom found in houses at this low price.

### FIRST FLOOR

**The Reception Hall.** On approaching the Verona you sense the feeling of welcome. The portico of colonial red brick set in white mortar, together with the dignified colonial front door with side lights, is the setting. Entering the reception hall is the colonial stairway of mahogany finish rail, white balusters, with stairs to match. A coat closet is located here for the accommodation of guests.

To the left and right of the hall are French doors, making it possible to open up the hall, living room and dining room into one for stately occasions or close them up at other times.

**The Living Room** extends the entire length of the house and is provided with a large cove or cornice mold extending around the ceiling, as shown in illustration on the opposite page. The colonial fireplace, together with the massive coves, go a long way to help furnish this beautiful room, which has eight windows and a pair of French doors leading to the large open porch. Size of living room, 13 feet 2 inches by 28 feet 9 inches.

**The Dining Room.** Size, 13 feet 2 inches by 15 feet 11 inches. Plenty of space for a full dining set, and for entertaining. It has four front windows and a double side window, insuring a cheerful atmosphere.

### Honor Bilt

## The Verona

No. P13201 "Already Cut" and Fitted

### $4,347.00

**The Kitchen** represents careful study. Size, 10 feet 8 inches by 12 feet 5 inches. The architect in locating the sink, range, refrigerator, kitchen table, folding ironing board, etc., had due consideration for the housewife, saving her many steps every day. The refrigerator compartment permits the filling of the refrigerator from the outside, keeping the iceman out of the kitchen. The patent ironing board disappears into the wall, giving the appearance of a panel door when closed. A study of the kitchen reveals:

**A Kitchen De Luxe Outfit** with its beautiful real tile sink, 11 feet long, swing stool and special cabinets on each side.

**Special built-in folding ironing board.**

**Refrigerator compartment, iced from outside.**

**Broom closet.**

**Breakfast alcove of spotless white.**

**The Breakfast Alcove** connected with the kitchen is a feature that is greatly appreciated by both housewife and servant. The alcove consists of two Dutch seats and a table to match. Here the morning breakfast can be served without disturbing the main dining room.

The lavatory, including the washstand and toilet, is located on the rear hall and convenient to reach from any room on the first floor.

### SECOND FLOOR

**The Bedrooms.** The colonial stairway in mahogany and white enamel is very attractive. This stairway leads directly to the second floor, permitting another stairway directly underneath which leads from the rear hall to the basement and very handy to the kitchen.

The second floor provides four good size bedrooms, each with cross ventilation, giving plenty of fresh air and light. The front bedroom is provided with triple mirror door closets, with drawers beneath for shoes; storage compartments with panel doors above. The other bedrooms have clothes closets provided with shelves and poles for clothes hangers.

Between the two front bedrooms is located the bathroom with a medicine chest with large plate glass mirror. Directly above the bathtub is a closet for towels, with three drawers below, which opens into hall.

**Attic.** Directly from the second floor hall is the stairway leading to the attic.

The rear porch provides an open balcony on the second floor to the French door leading to the stair landing. This is a convenient place for airing the bedding, etc.

**The Basement.** Room for heating plant, laundry, vegetable, fruit room, and storage.

**Height of Ceilings.** First floor, 9 feet from floor to ceiling. Second floor, 8 feet 2 inches from floor to ceiling. Basement, 7 feet from floor to joists.

### What Our Price Includes

At the price quoted we will furnish all the material to build this seven-room house, consisting of:
Lumber; Lath;
Roofing, Best Grade Clear Red Cedar Shingles;
Siding, Clear Cypress or Clear Red Cedar;
Framing Lumber, No. 1 Quality Douglas Fir or Pacific Coast Hemlock;
Flooring, Clear Oak for Living Room, Dining Room and Hall, First Floor; Clear Maple for Kitchen and Lavatory, Tile for Bathroom, Clear Douglas Fir or Pacific Coast Hemlock for all other rooms and halls;
Porch Flooring, Clear Edge Grain Fir for Side and Rear Porch;
Porch Ceiling, Clear Douglas Fir or Pacific Coast Hemlock;
Finishing Lumber;
High Grade Millwork (see pages 110 and 111);
Interior Doors, One-Panel Design for First Floor and Two-Panel Design for Second Floor, all of Douglas Fir;
Trim, Beautiful Grain Douglas Fir or Yellow Pine;
Windows of California Clear White Pine;
Medicine Case;                  Wardrobe;
Kitchen De Luxe Outfit;         Linen Case;
Kitchen Cupboard;               Flower Boxes;
Built-In Ironing Board;
Colonial Mantel;
Breakfast Alcove and Seats;
Colonial Shutters;
Eaves Trough and Down Spout;
40-Lb. Building Paper; Sash Weights;
Chicago Design Hardware (see page 132);
Paint for Three Coats Outside Trim and Siding;
Shellac, Varnish and Stain for Interior Doors, Stair Treads and Stair Rail; Enamel for all Interior Trim and Woodwork;
Shellac, Paste Filler and Floor Varnish for Oak and Maple Floors.
Complete Plans and Specifications.
We guarantee enough material to build this house. Price does not include cement, brick or plaster. See description of "Honor Bilt" Houses on pages 12 and 13.

### OPTIONS

*Sheet Plaster and Plaster Finish furnished in place of wood lath, $287.00 extra. See page 109.*
*Oriental Asphalt Shingles, guaranteed 17 years, instead of wood shingles, $53.00 extra.*
*Storm Doors and Windows, $137.00 extra.*
*Screen Doors and Windows, galvanized wire, $82.00 extra.*
*Seats for Porch, $29.00 per pair, extra.*
For prices of Plumbing, Heating, Wiring, Electric Fixtures and Shades see pages 130 and 131.

*For Our Easy Payment Plan See Page 144*

# The INTERIORS of the VERONA

Actual Photographs of Owner's House

*A BEDROOM*

*A BEDROOM*

*THE RECEPTION HALL*

*ABOVE RIGHT*—View of the Living Room

**FIRST FLOOR PLAN**

These Interiors were furnished by Ricaby and Co., Buffalo, N. Y., for the Verona, shown on the opposite page, which they built at Mt. Vernon on the Lake, a suburb of Buffalo, N. Y.

*THE BREAKFAST ALCOVE*

*LEFT*—Another View of the Living Room

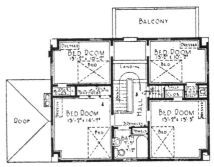

**SECOND FLOOR PLAN**

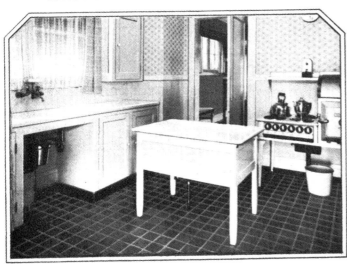

*LEFT* THE KITCHEN

*RIGHT* THE DINING ROOM

## The Sunbeam
### No. P3194A "Already Cut" and Fitted
### $2,707.00

THE SUNBEAM is a one story and a half modern bungalow of the better kind. Built in many exclusive suburbs, sometimes sold at handsome profits, and, at all times, the particular pride of its owners.

The Sunbeam is well planned, of "Honor Bilt" quality and construction, and a mighty good value whether bought for a home, an investment or for speculative purposes.

**The Exterior.** The Sunbeam has a large porch, protected by the main roof which is supported by four massive stuccoed pillars. The porch is 25 feet by 9 feet, has stairs at either end, and is ornamented by a long flower box. The open air sleeping porch, above, carries out the same decorative flower box ornament.

### FIRST FLOOR

**The Living Room** is spacious; 15 feet 2 inches by 17 feet. It is graced by a beautiful colonial mantel and fireplace. The open stairway to the second floor is located directly opposite the entrance. A coat closet for guests is near the stairway. Piano and furniture space is liberal, so that almost any kind of arrangement can be effected. Five windows provide a cheerful air.

**The Dining Room.** Buffet faces the wide cased opening to the living room. Size of dining room, 12 feet 8 inches by 11 feet 2 inches. Here, too, are five windows.

**The Kitchen** is modern in every detail. See floor plan for location of the built-in kitchen cabinet, and the broom closet providing storage for vacuum cleaner, brooms, mops, pails, scrub brushes, table leaves, etc. See the built-in folding ironing board; also space for the sink, range, table. A door leads to the rear entry which provides space for a refrigerator, handy to the iceman and away from the range. Stairs lead to grade door and to the basement.

### SECOND FLOOR

**The Bedrooms.** The stairway from the living room leads to a hall which connects with the two bedrooms, bathroom and linen closet.

The open air sleeping porch may be entered from either front bedroom. Bedrooms are well lighted and aired. Each bedroom has a clothes closet. The bath-room plumbing may be roughed-in on one wall, saving installation expenses. It has a medicine case.

**The Basement.** Room for furnace, laundry and storage.

**Height of Ceilings.** First floor, 9 feet from floor to ceiling. Second floor, 8 feet 2 inches from floor to ceiling. Basement, 7 feet from floor to joists.

### *What Our Price Includes*

**At the price quoted we will furnish all the material to build this five-room house, consisting of:**
Lumber; Lath;
**Roofing,** Oriental Asphalt Shingles, Guaranteed for 17 Years;
**Siding,** Clear Cypress or Clear Red Cedar Bevel, Clear Red Cedar Shingles Above Belt;
**Framing Lumber,** No. 1 Quality Douglas Fir or Pacific Coast Hemlock;
**Flooring,** Clear Maple in Kitchen and Bathroom, Clear Oak for Balance of First Floor, Clear Douglas Fir or Pacific Coast Hemlock for Balance of Second Floor;
**Porch Flooring,** Clear Edge Grain Fir;
**Porch Ceiling,** Clear Douglas Fir or Pacific Coast Hemlock;
**Finishing Lumber;**
**High Grade Millwork** (see pages 110 and 111);
**Interior Doors,** Two-Panel Design for First Floor and Five-Cross Panel Design for Second Floor of Douglas Fir;
**Trim,** Beautiful Grain Douglas Fir or Yellow Pine;
**Windows,** California Clear White Pine;
**Medicine Case;**
**Kitchen Cabinet;**
**Buffet;**
**Built-In Ironing Board;**
**Mantel;**
**Eaves Trough and Down Spouts;**
**40-Lb. Building Paper; Sash Weights;**
**Chicago Design Hardware** (see page 132);
**Paint** for Three Coats Outside Trim and Siding;
**Stain** for Shingles on Gable for Two Brush Coats;
**Shellac and Varnish** for Interior Trim and Doors;
**Shellac, Paste Filler and Floor Varnish** for Oak and Maple Floors;
**Screens** for Sleeping Porch.
Complete Plans and Specifications.

We guarantee enough material to build this house. Price does not include cement, brick or plaster. See description of "Honor Bilt" Houses on pages 12 and 13.

### OPTIONS

*We will furnish Sheet Plaster and Plaster Finish, instead of lath, for $173.00 extra. See page 109.*
*Storm Doors and Windows, $98.00 extra.*
*Screen Doors and Windows, galvanized wire, $46.00 extra.*

For prices of Plumbing, Heating, Wiring, Electric Fixtures and Shades see pages 130 and 131.

**Can be built on a lot 36 feet wide**

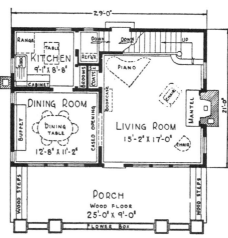

FIRST FLOOR PLAN

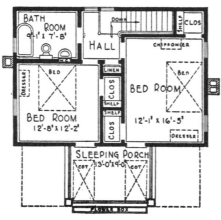

SECOND FLOOR PLAN

*For Our Easy Payment Plan See Page 144*

THE WESTLY is a high grade two-story home, retaining the architectural beauty of a modern bungalow. Built everywhere. Every customer satisfied. Praiseworthy letters from Westly owners tell of the fine interior arrangement, beautiful woodwork, our approved "Honor Bilt" ready-cut system of construction, and of savings even as high as $1,500.00.

**Exterior.** Sided with narrow bevel clear cypress siding in first story; dormer, roof and second story covered with best grade of thick cedar shingles. Large front porch, 30 feet by 8 feet. Porch can be screened or glazed.

**Can be built on a lot 35 feet wide**

**Honor Bilt**

## The Westly
### No. P13085 "Already Cut" and Fitted
## $2,614.00

### FIRST FLOOR

**The Living Room.** Size, 17 feet 8 inches by 13 feet 5 inches. An attractive feature is the open stairway that leads to the second floor. A coat closet with a mirror door is near the stairway. Furniture can be attractively arranged because of space. There are two windows at the side and one window at the front.

**The Music Room.** French doors connect with the living room. This music room is sometimes used for a bedroom instead. Size, 11 feet 2 inches by 9 feet 5 inches. Has a double window at the side and a high sash looking over the space for piano at the rear.

**The Dining Room.** A wide cased opening connects the living room and dining room. Size of dining room, 11 feet 2 inches by 12 feet 2 inches. Space for a complete dining set, including a buffet. Two side windows and one front window provide light and air.

**The Kitchen.** 11 feet 2 inches by 10 feet 8 inches. A swinging door connects with dining room. The space for sink, range, table and chair is laid out to save steps for the housewife. One side window and one rear window furnish light and cross ventilation. The pantry has five roomy shelves and a window.

A door from the kitchen opens into the rear porch, which has stairs to basement and to grade.

### SECOND FLOOR

**The Bedrooms.** Stairs from the living room lead directly into a well lighted hall. This hall connects with the three bedrooms, bathroom and linen closet. The three bedrooms are all of good size. The front bedroom has a door to balcony. Each bedroom has two windows and a spacious clothes closet with a window. One bedroom has two clothes closets.

**The Basement.** Space for furnace, laundry and storage.

**Height of Ceilings.** First floor, 9 feet from floor to ceiling. Second floor, 8 feet 2 inches from floor to ceiling. Basement, 7 feet from floor to joists.

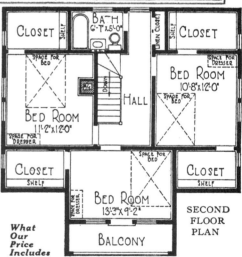

**SECOND FLOOR PLAN**

**What Our Price Includes**

At the price quoted we will furnish all the material to build this seven-room house, consisting of:
Lumber; Lath;
Roofing, Best Grade Clear Red Cedar Shingles;
Siding, Clear Cypress or Clear Red Cedar, Bevel;
Framing Lumber, No. 1 Quality Douglas Fir or Pacific Coast Hemlock;
Flooring, Clear Oak for Living Room, Dining Room and Music Room, Clear Maple for Kitchen, Pantry and Bathroom, Clear Douglas Fir or Pacific Coast Hemlock for Balance of Rooms;
Porch Flooring, Clear Edge Grain Fir;
Porch Ceiling, Clear Douglas Fir or Pacific Coast Hemlock;
Finishing Lumber;
High Grade Millwork (see pages 110 and 111);
Interior Doors, One-Panel Design for First Floor, and Two-Panel Design for Second Floor, All of Douglas Fir;
Trim, Beautiful Grain Douglas Fir or Yellow Pine;
Windows, California Clear White Pine;
Medicine Case;
Eaves Trough; Down Spout;
40-Lb. Building Paper; Sash Weights;
Chicago Design Hardware (see page 132);
Paint for Three Coats Outside Trim and Siding;
Stain for Two Brush Coats for Shingles on Gable Walls;
Shellac and Varnish for Interior Trim and Doors;
Shellac, Paste Filler and Floor Varnish for Oak and Maple Floors.
Complete Plans and Specifications.
We guarantee enough material to build this house. Price does not include cement, brick or plaster. See description of "Honor Bilt" Houses on pages 12 and 13.

### OPTIONS

Sheet Plaster and Plaster Finish, to take the place of wood lath, $234.00 extra. See page 109.
Oriental Asphalt Shingles, for roof, guaranteed 17 years, instead of wood shingles, $36.00 extra.
Oak Doors and Trim for living room, dining room, also Oak Stairs, $130.00 extra.
Storm Doors and Windows, $100.00 extra.
Screen Doors and Windows, galvanized wire, $62.00 extra.
For prices of Plumbing, Heating, Wiring, Electric Fixtures and Shades see pages 130 and 131.

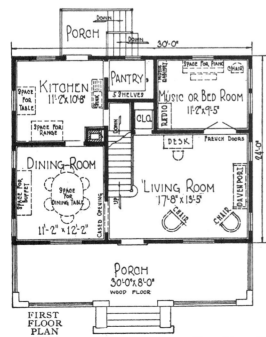

**FIRST FLOOR PLAN**

*For Our Easy Payment Plan See Page 144*

**Honor Bilt**

## The Kismet
### No. P17002 "Already Cut" and Fitted
### $1,097.00

IN THE KISMET bungalow we offer a good home at a low price with an absolute guarantee as to the "Honor Bilt" quality of the materials we furnish. This four-room bungalow is suitable for almost any location. In many sections a four or five-room cottage and lot will sell for $4,000.00 to $5,000.00. Notwithstanding the low price which we ask for all of the material required in the construction of this bungalow, there is no sacrifice of quality and there will be no shortage of material. As this house can be built on a lot 25 feet wide, it is suitable for town or country. For a farm house, for a small family, it represents a splendid investment. It is gracefully proportioned and when nicely painted will look well anywhere.

**INTERIOR VIEW, HOME OF A CUSTOMER**

**The Living and Dining Room.** From the shady porch, 15 feet by 6 feet, a glazed door opens into the modern combination living and dining room. Size of this room, 10 feet 8 inches by 17 feet 8 inches. The front part is used for living quarters and the part near the kitchen door is used for dining.

**The Kitchen** is 8 feet 8 inches by 7 feet 2 inches. It has space for sink, range, table and chair; also a closet with shelves. A double window provides light and ventilation. Immediately outside the kitchen door, in the rear entry, is space for the ice box. Here stairs lead to grade door and basement.

**The Bedrooms.** The front bedroom opens from the living and dining room. The rear bedroom and the bathroom are entered from an open hall, right off the dining end of room. Each bedroom has a clothes closet, and is lighted and aired by a window.

**The Basement.** Room for furnace, laundry and storage.

**Height of Ceilings.** Main floor, 9 feet from floor to ceiling. Basement, 7 feet from floor to joists.

### What Our Price Includes

At the price quoted we will furnish all the material to build this four-room bungalow, consisting of:

**Lumber; Lath;**
**Roofing,** Best Grade Clear Red Cedar Shingles;
**Siding,** Clear Cypress or Clear Red Cedar, Bevel;
**Framing Lumber,** No. 1 Quality Douglas Fir or Pacific Coast Hemlock;
**Flooring,** Clear Douglas Fir or Pacific Coast Hemlock;
**Porch Flooring,** Clear Edge Grain Fir;
**Porch Ceiling,** Clear Douglas Fir or Pacific Coast Hemlock;
**Finishing Lumber;**
**High Grade Millwork** (see pages 110 and 111);
**Interior Doors,** Five-Cross Panel Design of Douglas Fir;
**Trim,** Beautiful Grain Douglas Fir or Yellow Pine;
**Windows,** California Clear White Pine;
**Medicine Case;**
**Eaves Trough and Down Spout;**
**40-Lb. Building Paper; Sash Weights;**
**Stratford Design Hardware** (see page 132);
**Paint** for Three Coats Outside Trim and Siding;
**Shellac and Varnish** for Interior Trim and Doors.
  Complete Plans and Specifications.
  We guarantee enough material to build this house. Price does not include cement, brick or plaster. See description of "Honor Bilt" Houses on pages 12 and 13.

**Can be built on a lot 25 feet wide**

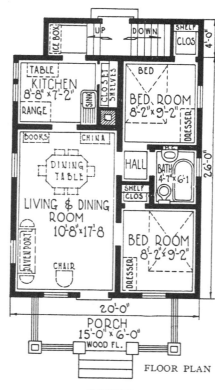

**FLOOR PLAN**

### OPTIONS

*Sheet Plaster and Plaster Finish, $114.00 extra. See page 109.*

*Oriental Asphalt Shingles, guaranteed 17 years, instead of wood shingles, $28.00 extra.*

*Storm Doors and Windows, $36.00 extra.*

*Screen Doors and Windows, galvanized wire, $25.00 extra.*

For prices of Plumbing, Heating, Wiring, Electric Fixtures and Shades see pages 130 and 131.

*For Our Easy Payment Plan See Page 144*

THE JOSEPHINE is the choice of many families because: (1) It has four good size rooms, large closets, bath and a shady porch. (2) It is built almost square, and therefore has no waste space, and is erected without waste of material. (3) It is of "Honor Bilt" quality and construction throughout. (4) It is priced low, since there are no middlemen's profits.

**The Exterior.** Sided with narrow bevel cypress siding. Gable over the porch to be sided with stucco on metal lath. The porch measures 14 feet 2 inches by 8 feet 2 inches.

**Can be built on a lot 34 feet wide**

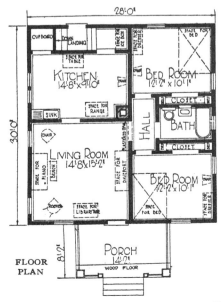

FLOOR PLAN

**Honor Bilt**

## The Josephine
No. P7044 "Honor Bilt" "Already Cut" and fitted.
### $1,452.00

**The Living Room.** You enter the large almost square, living room from the porch. Size, 14 feet 8 inches by 15 feet 2 inches. Long wall space allows room for a piano, davenport, etc. There is plenty of light, for the porch roof does not cover the front window, the door is glazed, and two small sash let in light on the side.

The living room can also be used as a dining room, if the kitchen is not used for this purpose.

**The Kitchen** is planned so that the work can all be done quickly in one part of the room. The large cupboard will hold dishes, pans and supplies. Space for refrigerator is on the landing just outside the kitchen door, made light by the glass in the outside door. The kitchen measures 14 feet 8 inches by 9 feet 10 inches.

**The Bedrooms.** From the living room a plastered opening leads to a hall that connects with the two bedrooms and bathroom. You seldom see such nice large bedrooms in a house of this size. Each of these bedrooms has an 8-foot closet with doors so arranged that every part of the closet can be easily reached. Each bedroom has two windows.

**The Basement.** Room for furnace, laundry and storage.

**Height of Ceilings.** Main floor, 9 feet from floor to ceiling. Basement, 7 feet from floor to joists.

### What Our Price Includes

At the price quoted we will furnish all the material to build this four-room house, consisting of:

Lumber; Lath;
**Roofing**, Best Grade of Clear Red Cedar Shingles;
**Siding**, Clear Cypress or Clear Red Cedar, Bevel;
**Framing Lumber**, No. 1 Quality Douglas Fir or Pacific Coast Hemlock;
**Flooring**, Clear Douglas Fir or Pacific Coast Hemlock;
**Porch Flooring**, Clear Edge Grain Fir;
**Porch Ceiling**, Clear Douglas Fir or Pacific Coast Hemlock;
**Finishing Lumber**;
**High Grade Millwork**, (see pages 110 and 111);
**Interior Doors**, Five Cross Panel Design of Douglas Fir;
**Trim**, Beautiful Grain Douglas Fir or Yellow Pine;
**Windows**, California Clear White Pine;
**Medicine Case**;
**Eaves Trough; Down Spouts**;
**40-Lb. Building Paper; Sash Weights**;
**Stratford Design Hardware** (see page 132);
**Paint** for Three Coats Outside Trim and Siding;
**Shellac and Varnish** for Interior Trim and Doors.

Complete Plans and Specifications.

We guarantee enough material to build this house. Price does not include cement, brick or plaster. See description of "Honor Bilt" Houses on pages 12 and 13.

### OPTIONS

*Sheet Plaster and Plaster Finish to take the place of wood lath, $140.00 extra. See page 109.*
*Oriental Asphalt Shingles, guaranteed 17 years, instead of wood shingles, $32.00 extra.*
*Storm Doors and Windows, $45.00 extra.*
*Screen Doors and Windows, galvanized wire, $30.00 extra.*

For prices of Plumbing, Heating, Wiring, Electric Fixtures and Shades, see pages 130 and 131.

*For Our Easy Payment Plan See Page 144*

**Honor Bilt**

## The Somers
### No. P17008 "Already Cut" and Fitted
### $1,778.00

THE SOMERS is an attractive and inexpensive bungalow of five rooms with vestibule, bath, and built-in cupboard. It has a front porch across the entire house, so arranged that it is easily converted to a sun room by glazing the open parts. The porch measures 22 feet 6 inches by 8 feet.

**The Living Room.** The front door leads from the porch into the vestibule which has a cased opening into the living room. Size of living room, 17 feet 8 inches by 12 feet 2 inches. Furniture and piano can be grouped to please because of the wall and floor space. Three front windows, two high sash, on the side, provide an abundance of light and air.

**The Dining Room** and the living room are practically one large room, as they are combined by the wide cased opening. The dining room has a recessed bay with three windows. Size, 12 feet 4 inches by 14 feet.

**The Kitchen.** A swinging door from the dining room leads to the kitchen that is lighted by two windows. Floor area of the kitchen is 8 feet by 12 feet 9 inches. There is space for the sink, the range, the table and chair. The built-in cupboard is located opposite the space for range. Stairs to the basement are through a door. Another door leads to porch and grade.

**The Bedrooms.** Directly off the dining room is an open hall that connects with the two bedrooms and bathroom. Each bedroom has a clothes closet, and is lighted by a window.

**The Bathroom** has a medicine case, and is lighted by a window. All of the plumbing can be roughed-in on one wall, saving installation expense.

**The Basement.** Room for furnace, laundry and storage.

**Height of Ceilings.** Main floor, 8 feet 6 in. from floor to ceiling. Basement, 7 feet from floor to joists.

The Somers has been built at Alden, Pa., Park Ridge, Ill., Cleveland, Ohio, Moscow, Kan., Kenosha, Wis., Alamogordo, N. M., Maxwell, Neb., and other cities.

### What Our Price Includes

At the price quoted we will furnish all the material to build this five-room bungalow, consisting of:

Lumber; Lath;
Roofing, Best Grade Clear Red Cedar Shingles;
Siding, Best Grade Clear Red Cedar Shingles;
Framing Lumber, No. 1 Quality Douglas Fir or Pacific Coast Hemlock;
Flooring, Clear Douglas Fir or Pacific Coast Hemlock;
Porch Flooring, Clear Edge Grain Fir;
Porch Ceiling, Clear Douglas Fir or Pacific Coast Hemlock;
Finishing Lumber;
High Grade Millwork (see pages 110 and 111);
Interior Doors, Five Cross Panel Design of Douglas Fir;
Trim, Beautiful Grain Douglas Fir or Yellow Pine;
Windows, California Clear White Pine;
Medicine Case;
Kitchen Cupboards;
Eaves Trough and Down Spouts;
40-Lb. Building Paper; Sash Weights;
Stratford Design Hardware (see page 132);
Paint for Three Coats Outside Trim;
Stain for Shingles on Walls for Two Brush Coats;
Shellac and Varnish for Interior Trim and Doors.

Complete Plans and Specifications.

We guarantee enough material to build this house. Price does not include cement, brick or plaster. See description of "Honor Bilt" Houses on pages 12 and 13.

### OPTIONS

*Sheet Plaster and Plaster Finish to take the place of wood lath, $148.00 extra. See page 109.*

*Oriental Asphalt Shingles, guaranteed 17 years, instead of wood shingles, for roof, $46.00 extra.*

*Oak Doors, Trim and Floor in vestibule, living room and dining room; Maple Floors in kitchen and bathroom, $158.00 extra.*

*Storm Doors and Windows, $51.00 extra.*

*Screen Doors and Windows, galvanized wire, $37.00 extra.*

For prices of Plumbing, Heating, Wiring, Electric Fixtures and Shades see pages 130 and 131.

**Can be built on a lot 30 feet wide**

This house can be built with rooms reversed. (See page 3.)

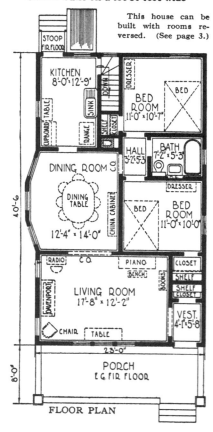

**FLOOR PLAN**

HERE is one of the many letters we have from Somers owners:

Sears, Roebuck and Co

Gentlemen:—Enclosed you will find a photo of our new bungalow built from your material and plans (with a few slight changes). I wish to say that all the material was extra fine. I saved at least $300.00 by buying from you. Thanking you for your honest treatment in these transactions, I am,
Yours respectfully,
CHARLES MILLER.
P. O. Box B, West Main St., Parkesburg, Pa.

### For Our Easy Payment Plan See Page 144

A CUSTOMER who built The Kilbourne bungalow recently wrote us as follows: "Our house has been the object of much admiration, not only from our friends, but strangers, who in passing by will stop to look at the artistic front. Many have remarked about the 'homey' porch. We have no hesitancy whatever in recommending Sears-Roebuck lumber, which came in plenty of time, and in splendid condition. Also must thank you for the courteous treatment and helpful suggestions you have given us. We know we saved nothing less than $1,500.00."

The Kilbourne bungalow satisfies every family that has built it. Judge for yourself! The photograph and floor plan reproduced on this page shows the reason why The Kilbourne is such an outstanding value. See its sloping roof, the dormer, the overhanging eaves, the fireplace chimney, the large porch and the massive porch pillars!

**The Living Room.** Size, 21 feet by 13 feet 2 inches. Interest is centered on the fireplace and mantel, at each side of which is a window. There are three additional windows overlooking the front lawn. The large size of this room allows for a pleasing arrangement of furnishings.

**Can be built on a lot 45 feet wide.**

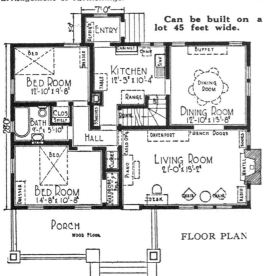

FLOOR PLAN

**Honor Bilt**

## The Kilbourne
### No. P17013 "Already Cut" and Fitted
## $2,700.00

**The Dining Room.** French doors connect the living room and dining room. Floor size of the dining room, 12 feet 10 inches by 13 feet 8 inches, just the right size for the modern home. A double side window and two high sash windows provide light and air.

**The Kitchen.** From the dining room a swinging door opens into the kitchen. Size of kitchen, 12 feet 3 inches by 10 feet 4 inches. It has a built-in cabinet, Nos. P9260 and P9261 shown on pages 110 and 111, space for sink, range, table and chairs. A double window affords light and ventilation.

In one corner of the kitchen there are five shelves, and on the opposite side a door opens to stairway leading down to the basement. At another end a door opens to stairway leading to the second floor. A door leads to the rear entry, which has space for a refrigerator, and door to grade stairs.

**The Bedrooms.** A hall connects the living room, the kitchen, the two bedrooms, the bathroom, and the hall coat closet. The front bedroom, 14 feet 8 inches by 10 feet 8 inches, has a big combination wardrobe, No. P9265, as illustrated on pages 110 and 111. Two front windows and one side window provide light and cross current of air. The rear bedroom, 12 feet by 9 feet 8 inches, has a clothes closet, and a window on each outer wall.

**The Bathroom** has a built-in medicine case.

**The Basement.** Room for furnace, laundry and storage.

**Height of Ceilings.** First floor, 9 feet from floor to ceiling. Basement, 7 feet from floor to joists.

### What Our Price Includes

At the price quoted we will furnish all the material to build this five-room bungalow, consisting of:

Lumber; Lath;
Roofing, Best Grade Clear Red Cedar Shingles;
Siding, Clear Cypress or Clear Red Cedar, Bevel. Best Grade of Clear Red Cedar Shingles on Porch Gable Wall;
Framing Lumber, No. 1 Quality Douglas Fir or Pacific Coast Hemlock;
Flooring, Clear Oak and Maple;
Porch Flooring, Clear Edge Grain Fir;
Porch Ceiling, Clear Douglas Fir or Pacific Coast Hemlock;
Finishing Lumber;
High Grade Millwork (see pages 110 and 111);
Interior Doors, Two-Panel Design of Douglas Fir;
Trim, Beautiful Grain Douglas Fir or Yellow Pine;
Windows, California Clear White Pine;
Medicine Case; Wardrobe;
Kitchen Cabinet; Brick Mantel;
Eaves Trough and Down Spouts;

ATTIC FLOOR PLAN
See Options

40-Lb. Building Paper; Sash Weights;
Chicago Design Hardware (see page 132);
Paint for Three Coats outside Trim and Siding;
Stain for Two Brush Coats for Shingles on Porch Gable Wall;
Shellac and Varnish for Interior Trim and Doors;
Shellac, Paste Filler and Floor Varnish for Oak and Maple Floors.
Complete Plans and Specifications.
We guarantee enough material to build this house. Price does not include cement, brick or plaster. See description of "Honor Bilt" Houses on pages 12 and 13.

### OPTIONS

*Furnished with three rooms in attic, with single floor, $300.00 extra. See attic plan above.*
*Sheet Plaster and Plaster Finish to take the place of wood lath, first floor, $182.00 extra; for first floor and attic, $293.00 extra. See page 109.*
*Oriental Asphalt Shingles, guaranteed 17 years, instead of wood shingles, $57.00 extra.*
*Oak Doors and Trim for living room and dining room, $84.00 extra.*
*Storm Doors and Windows, $73.00 extra; with attic, $100.00 extra.*
*Screen Doors and Windows, galvanized wire, $46.00 extra; with attic, $63.00 extra.*

For prices of Plumbing, Heating, Wiring, Electric Fixtures and Shades see pages 130 and 131.

View of front bedroom showing wardrobe.

*For Our Easy Payment Plan See Page 144*

A NEAT home with five comfortable rooms and bath, conservative and economical. The wonder of the house is how exactly right all the little points are. Grade entrance; space for ice box on same level as kitchen floor; front bedroom and kitchen have cross ventilation; bathroom is entered from a little hallway and not directly from living room, bathroom unusually well arranged; three big closets and all rooms well planned to accommodate furniture.

### No. P13052

**The Living Room.** The large living room and cheerful dining room are connected by means of a wide cased opening. Opposite the front door with windows grouped about it is a long space against the wall where any handsome large piece of furniture will look attractive. In fact, the living room is one that will be easy to furnish cozily. Size, 13 feet 5 inches by 13 feet 5 inches.

**The Dining Room.** Entered from the living room through a wide cased opening. Measures 10 feet 10 inches by 12 feet 5 inches. Has good space for the dining set and is lighted from two sides.

**Can be built on a lot 42 feet wide**

### Honor Bilt
## The Conway

| | |
|---|---|
| No. P13052 "Already Cut" and Fitted | $1,707.00 |
| No. P3052 "Already Cut" and Fitted | 1,613.00 |

**The Kitchen.** Right at the swinging door from the dining room is the sink with a window beside it, and close by is space for the range. Under the window on the opposite wall a table can be set. Opening the rear door you find a landing where the ice box can stand and just inside this door is a good sized shelf where things to be put into the refrigerator can be gathered and all put in at one time, saving ice. The shelf will be useful in many ways, and when cleaning the kitchen it can be dropped flat against the wall. Size of the kitchen, 9 feet 4 inches by 10 feet 5 inches.

**The Bedrooms.** The front bedroom opens up from the living room. The rear bedroom connects with an open hall off the living room. Each bedroom has a clothes closet, and is well lighted and aired. The bathroom connects with the hall, also. It has a medicine case.

### THE ATTIC

**Two additional rooms** may be obtained at a slight extra cost. See options.

**The Basement.** Room for furnace, laundry and storage.

**Height of Ceilings.** First floor, 9 feet from floor to ceiling. Basement, 7 feet from floor to joists.

### No. P3052

Same as No. P13052, except that house is two feet narrower. See floor plan to right.

### What Our Price Includes

At the price quoted we will furnish all the material to build this five-room house consisting of:
Lumber; Lath;
Roofing, Clear Red Cedar Shingles;
Siding, Clear Cypress, Bevel;
Framing Lumber, No. 1 Quality Douglas Fir or Pacific Coast Hemlock;
Flooring, Clear Douglas Fir or Pacific Coast Hemlock;
Porch Flooring, Clear Edge Grain Fir;
Porch Ceiling, Clear Douglas Fir or Pacific Coast Hemlock;
Finishing Lumber;
High Grade Millwork (see pages 110 and 111);
Interior Doors, Five Cross Panel Design of Douglas Fir;
Trim, Beautiful Grain Douglas Fir;
Windows of California Clear White Pine;
Medicine Case;
Eaves Trough and Down Spout;
40-Lb. Building Paper; Sash Weights;
Stratford Design Hardware (see page 132);
Paint for Three Coats Outside, Trim and Siding;
Shellac and Varnish for Interior Doors and Trim.
Complete Plans and Specifications.
We guarantee enough material to build this house.
Price does not include cement, brick or plaster.
See description of "Honor Bilt" Houses on pages 12 and 13.

#### OPTIONS

*Furnished with two rooms in the attic, with single floor, $196.00 extra for No. P3052; $229.00 extra for No. P13052.*
*Sheet Plaster and Plaster Finish to take the place of wood lath, $142.00 extra; with attic, $195.00 extra for No. P3052; $149.00 extra, with attic, $201.00 extra for No. P13052. See page 109.*
*Oriental Asphalt Shingles, guaranteed 17 years, instead of wood shingles, $38.00 extra for No. P3052; $42.00 extra for No. P13052.*
*Oak Trim, Doors and Floors in living and dining rooms, $136.00 extra.*
*Storm Doors and Windows for first floor, $46.00 extra; with attic, $58.00 extra.*
*Screen Doors and Windows, galvanized wire, for first floor, $30.00 extra; with attic, $37.00 extra.*
For prices on Plumbing, Heating, Wiring, Electric Fixtures and Shades, see pages 130 and 131.

**Can be built on a lot 40 feet wide**

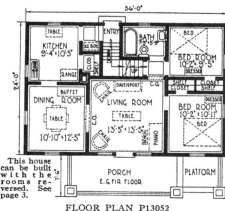

This house can be built with the rooms reversed. See page 3.

FLOOR PLAN P13052

**The Conway
Front and Side View**

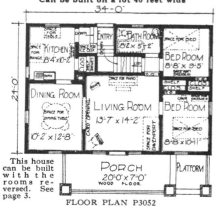

This house can be built with the rooms reversed. See page 3.

FLOOR PLAN P3052

*For Our Easy Payment Plan See Page 144*

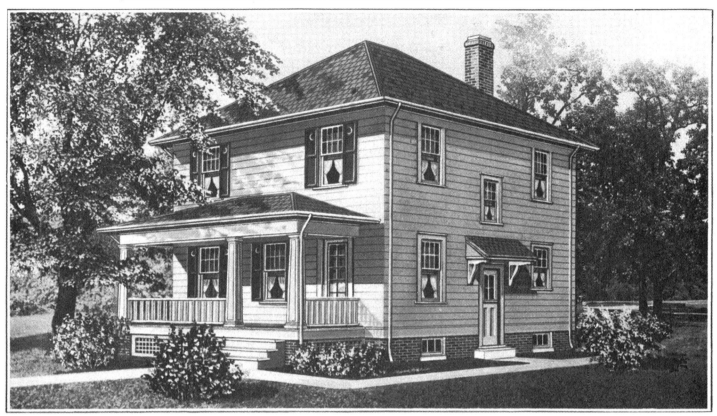

THE ALBION two-story residence has attractive qualities which reflect the most modern trend of architecture. These qualities include utmost use of material and greatly lowered construction cost because the building is almost square. Every inch of floor space is available for living quarters.

Painted pure white with contrasting green shutters and red or green roof and red brick chimney. Consider the large porch, 20 feet by 8 feet, with its artistic columns; the numerous, divided light windows and the wide bevel siding. Just study the floor plans and see for yourself how perfect in arrangement is the Albion "Honor Bilt" Home. Our direct-from-factory-to-you system makes the price low. There is a strong demand for this kind of a home, and handsome profits are the rule for our customers.

### FIRST FLOOR

**The Living Room.** The front door, with its ten lights, opens directly into the vestibule. Directly ahead a stairway leads to the second floor. To the left of stairway is a coat closet with a mirror door. And here, too, a wide cased

**Can be built on a lot 32 feet wide**

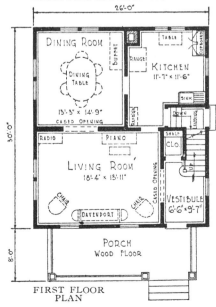

FIRST FLOOR
PLAN

**Honor Bilt**

## The Albion

No. P3227 "Already Cut" and Fitted
### $2,515.00

opening invites the guest to the handsome and spacious living room. Size, 18 feet 4 inches by 13 feet 11 inches, an ideal size. It has ample wall space for a piano, davenport and other furniture. Two front windows and a double window at the side provide sunshine and air.

**The Dining Room.** A wide cased opening between the living room and dining room makes both rooms available as one extra large room when entertaining. Size of the dining room, 13 feet 3 inches by 14 feet 9 inches. Here, again, liberal space allows for a complete dining set, including a buffet. There is a double window at the side and another double window at the rear.

**The Kitchen.** From the dining room a swinging door connects with the bright kitchen. Size, 11 feet 7 inches by 11 feet 6 inches. The housewife happily attends to her daily tasks amidst modern service features: There is a built-in kitchen cabinet. Space for refrigerator is close to the door which opens to side entry. Another door connects with living room, saving steps when answering the front door bell. The side entry leads to the basement and to the outside.

### SECOND FLOOR

**The Bedrooms.** A hall connects with the bedrooms, bathroom, and linen closet. The hall is lighted by a window on the stair landing. Each of the four bedrooms has a good size clothes closet. There are two windows in each bedroom, one window on each outside wall, thereby assuring permanent light and cross ventilation.

**The Bathroom** is fitted with a medicine case. It has one window.

**Basement.** Room for furnace, laundry and storage.

**Height of Ceilings.** First floor, 9 feet from floor to ceiling. Second floor, 8 feet 6 inches from floor to ceiling. Basement, 7 feet from floor to joists.

### What Our Price Includes

At the price quoted we will furnish all the material to build this seven-room house, consisting of:
Lumber; Lath;
Roofing, Oriental Slate Surfaced Shingles, Guaranteed 17 Years;
Siding, Clear Cypress or Red Cedar, Bevel;
Framing Lumber, No. 1 Quality Douglas Fir or Pacific Coast Hemlock;
Flooring, Clear Grade Douglas Fir or Pacific Coast Hemlock;
Porch Flooring, Clear Edge Grain Fir;
Porch Ceiling, Clear Grade Douglas Fir or Pacific Coast Hemlock;

**Can be built on a lot 32 feet wide**

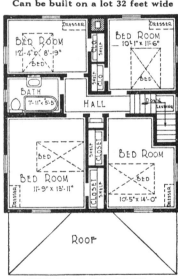

SECOND FLOOR PLAN

Finishing Lumber;
High Grade Millwork (see pages 110 and 111);
Interior Doors, Two-Cross Panel Design of Douglas Fir;
Trim, Beautiful Grain Douglas Fir or Yellow Pine;
Windows, California Clear White Pine;
Medicine Case;
Kitchen Cupboards;
Colonial Shutters;
Eaves Trough and Down Spouts;
40-Lb. Building Paper; Sash Weights;
Chicago Design Hardware (see page 132);
Paint for Three Coats Outside Trim and Siding;
Shellac and Varnish for Interior Trim and Doors.
Complete Plans and Specifications.
We guarantee enough material to build this house. Price does not include cement, brick or plaster.
See description of "Honor Bilt" Houses on pages 12 and 13.

### OPTIONS

*Sheet Plaster and Plaster Finish to take the place of wood lath, $250.00 extra. See page 109.*

*Oak Doors, Trim and Floors for living room, dining room, and reception hall, also Oak Stairs; Maple Floors in kitchen and bathroom, $216.00 extra.*

*Storm Doors and Windows, $82.00 extra.*

*Screen Doors and Windows, galvanized wire, $52.00 extra.*

For prices of Plumbing, Heating, Wiring, Electric Fixtures and Shades see pages 130 and 131.

*For Our Easy Payment Plan See Page 144*

THE WOODLAND is of the type of architecture that meets with favor wherever it is built. Besides having nine rooms, it is planned with a large front porch, 26 feet wide by over 8 feet deep. A study of the picture reveals many attractive features—the wide steps leading to the porch, the triple columns on either side of the porch, supporting an ornamental truss for the porch roof. All rafter ends on porch and main roof have exposed ends. The front windows have divided upper lights. The two front windows on the second story are provided with flower boxes. Even the dormer adds to the symmetry of a harmonious design that is difficult to improve. Being planned on strictly square lines, every foot of floor space is utilized to advantage and the upkeep is small.
**Can be built on a lot 32 feet wide.**

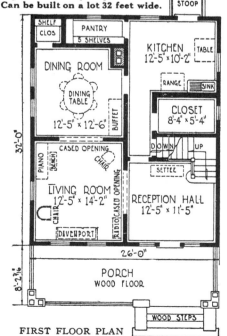

**FIRST FLOOR PLAN**

**Honor Bilt**

## The Woodland
### No. P3025 "Already Cut" and Fitted
## $2,491.00

### FIRST FLOOR

**The Living Room.** The front door, glazed with a large light, and with sash on either side, makes an imposing entrance, and furnishes plenty of sunshine and air to the large reception hall, 12 feet 5 inches by 11 feet 5 inches, in which is located the stairway to the second floor. As a wide cased opening is provided between this hall and the living room, which is 12 feet 5 inches by 14 feet 2 inches, the combined space gives the appearance of one large living room, thus allowing sufficient room for furniture, including piano.

**The Dining Room.** Size, 12 feet 5 inches by 12 feet 6 inches, is directly back of the living room. It is connected by a large cased opening, thus throwing the three rooms into one, when the occasion warrants it. Here is space for buffet or china closet, with plenty of room for table and chairs. There is also a closet for linen, etc.

**The Kitchen.** It is directly to the right of dining room, permits a handy arrangement of range, sink and work table, and connects with large pantry in the rear, and hallway to the reception hall in front.

A closet, 8 feet 4 inches by 5 feet 4 inches, is located directly off the first floor hall. This room can be used for lavatory, sewing room or closet.

A stairway, directly under the main stairway, leads to the basement.

### SECOND FLOOR

**The Bedrooms.** The stairway leads to hall on second floor, that has direct connection with five bedrooms and bath. There are four clothes closets on this floor.

**The Basement.** Excavated basement with concrete floor. Room for furnace, laundry and storage.

**Height of Ceilings.** First floor, 9 feet from floor to ceiling. Second floor, 8 feet 2 inches from floor to ceiling. Basement, 7 feet from floor to joist.

### What Our Price Includes

At the price quoted we will furnish all the material to build this nine-room house, consisting of:
Lumber; Lath;
Roofing, Best Grade Clear Red Cedar Shingles;
Siding, Clear Cypress or Clear Red Cedar, Bevel;
Framing Lumber, No. 1 Quality Douglas Fir or Pacific Coast Hemlock;
Flooring, Clear Douglas Fir or Pacific Coast Hemlock;
Porch Flooring, Clear Edge Grain Fir;

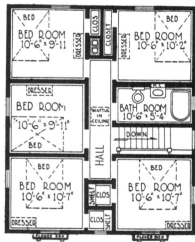

**SECOND FLOOR PLAN**

Porch Ceiling, Clear Douglas Fir or Pacific Coast Hemlock;
Finishing Lumber;
High Grade Millwork (see pages 110 and 111);
Interior Doors, Five Cross Panel Design of Douglas Fir;
Trim, Beautiful Grain Douglas Fir or Yellow Pine;
Medicine Case;
Windows, California Clear White Pine;
40-Lb. Building Paper; Sash Weights;
Eaves Trough and Down Spout;
Chicago Design Hardware (see page 132);
Paint for Three Coats Outside Trim and Siding;
Stain for Two Brush Coats for Shingles on Dormer Walls;
Shellac and Varnish for Interior Trim and Doors;
Complete Plans and Specifications.

Built on concrete foundation and excavated under entire house.

We guarantee enough material to build this house. Price does not include cement, brick or plaster.

See description of "Honor Bilt" Houses on pages 12 and 13.

### OPTIONS

*Sheet Plaster and Plaster Finish, to take the place of wood lath, $288.00 extra. See page 109.*

*Oriental Asphalt Shingles, guaranteed 17 years, instead of wood shingles, $35.00 extra.*

*Oak Doors, Trim and Floors for reception hall, living room and dining room; also Oak Stairs. Maple Floors in kitchen and bathroom, $202.00 extra.*

*Storm Doors and Windows, $80.00 extra.*

*Screen Doors and Windows, galvanized wire, $50.00 extra.*

For prices of Plumbing, Heating, Wiring, Electric Fixtures and Shades see pages 130 and 131.

*For Our Easy Payment Plan See Page 144*

**Can be built on a lot 28 feet wide**

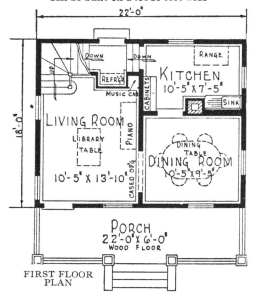

FIRST FLOOR PLAN

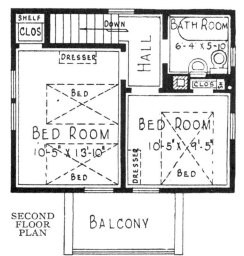

SECOND FLOOR PLAN

## The Windsor

No. P3193 "Already Cut" and Fitted

### $1,577.00

T HE WINDSOR is a two-story home, styled in the popular semi-bungalow architecture. Every detail of its beautiful exterior, and every room within, is planned to give lasting satisfaction.

The low price of the Windsor is proof of the big saving resulting from our "Honor Bilt" system. Besides, our direct-from-the-factory-to-you method has no equal.

**Exterior.** Just observe the striking effect of the second floor balcony! Also, the purlins underneath the wide overhanging eaves, the spacious front porch, 22 feet by 6 feet, and the combination of shingles and cypress siding!

### FIRST FLOOR

**The Living Room** is entered from the front porch. It measures 10 feet 5 inches by 13 feet 10 inches. Directly opposite the entrance is the open stairway to the second floor. A piano and the furniture may be grouped attractively. Near the space for the piano is a built-in music cabinet. Light and air from two sides.

**The Dining Room** and the living room are connected by a wide cased opening. Size of dining room is 10 feet 5 inches by 9 feet 5 inches. Here the dining set has plenty of floor space, and accommodation for family and guests.

**The Kitchen.** From the dining room a swinging door leads to the bright and cheery kitchen. The built-in kitchen cabinet and the space for sink, range and table is handy, so that time and effort is not wasted. A door leads to the rear entry which has space for a refrigerator, stairs to basement, and grade.

### SECOND FLOOR

**The Bedrooms.** A stairway from the living room leads to the hall on the second floor. This hall connects with the two bedrooms and the bathroom. Each bedroom has a clothes closet, and is well lighted and ventilated by windows.

**The Basement.** Room for furnace, laundry and storage.

**Height of Ceilings.** First floor, 9 feet from floor to ceiling. Second floor, 8 feet 2 inches from floor to ceiling. Basement, 7 feet from floor to joists.

### What Our Price Includes

At the price quoted we will furnish all the material to build this five-room house, consisting of:

**Lumber; Lath;**
**Roofing,** Best Grade Clear Red Cedar Shingles;
**Siding,** Clear Cypress or Clear Red Cedar Bevel Below Belt Course, Best Grade Clear Red Cedar Shingles Above Belt Course;
**Framing Lumber,** No. 1 Quality Douglas Fir or Pacific Coast Hemlock;
**Flooring,** Clear Douglas Fir or Pacific Coast Hemlock;
**Porch Flooring,** Clear Edge Grain Fir;
**Porch Ceiling,** Clear Douglas Fir or Pacific Coast Hemlock;
**Finishing Lumber;**
**High Grade Millwork** (see pages 110 and 111);
**Interior Doors,** Five Cross Panel Design of Douglas Fir;
**Trim,** Beautiful Grain Douglas Fir or Yellow Pine;
**Windows,** California Clear White Pine;
**Medicine Case;**
**Kitchen Cabinets;**
**Eaves Trough and Down Spout;**
**40-Lb. Building Paper; Sash Weights;**
**Stratford Design Hardware** (see page 132);
**Paint** for Three Coats Outside Trim and Siding;
**Stain** for Two Brush Coats for Shingles on Walls;
**Shellac and Varnish** for Interior Trim and Doors.
Complete Plans and Specifications.

We guarantee enough material to build this house. Price does not include cement, brick or plaster. See description of "Honor Bilt" Houses on pages 12 and 13.

### OPTIONS

*Sheet Plaster and Plaster Finish, to take the place of wood lath, $128.00 extra. See page 109.*
*Oriental Asphalt Shingles, guaranteed 17 years, instead of wood shingles, $19.00 extra.*
*Storm Doors and Windows, $57.00 extra.*
*Screen Doors and Windows, galvanized wire, $37.00 extra.*

For prices of Plumbing, Heating, Wiring, Electric Fixtures and Shades see pages 130 and 131.

**For Our Easy Payment Plan See Page 144**

THE OSBORN is the most pleasing type of stucco and shingle sided bungalow in Spanish mission architecture. Where will you find its equal? Massive stucco porches and bulkheads, trimmed with red brick coping, give that needed touch of color, emphasizing its graceful lines. The timber columns resting on the large square piers or concrete columns, are in perfect harmony with the rest, and support the graceful roof with its wide verge boards and timber purlins. Here the architect has given careful study to every detail, and furnished a creation that is striking, yet restful. The shingle siding, the timber wood columns, corbels and purlins, can be painted or stained a rich brown, or dark brick red, with most pleasing effect, in contrast with the gray stucco porch walls and chimney.

The Osborn will appeal to the lover of nature because of its two open porches both sheltered by the main roof, and the sleeping porch in the rear. The side porch is private, size 13 feet by 8 feet 8 inches. The front porch is provided with steps and landing leading to the front entrance, and is 22 feet by 9 feet, which is of unusual size.

**The Living Room.** This spacious room extends the entire width of the house. Size, 23 feet 11 inches by 11 feet 11 inches. One is immediately impressed with the cozy brick mantel and fireplace which is at one end of the room. At each side of mantel is a bookcase with leaded glass doors, and a nine-light window above each bookcase. A window seat with hinged cover is at the other end of the room. The window seat has storage space for clothing, blankets, etc. At each side of the window seat is a coat closet. Ample wall space will accommodate a piano and pictures, while the large floor area provides a good setting for furniture, including davenport. Five windows furnish an abundance of light and ventilation.

**The Dining Room.** A cased opening with a bookcase on either side, leads from the living room to the dining room. Size, 13 feet 7 inches by 15 feet 1 inch. Walls are paneled according to the latest mode. Space accommodates a complete dining room set. French doors lead to private side porch. A cheerful atmosphere is assured with a flood of light and air from the windows and French doors.

# The Osborn
### No. P12050A "Already Cut" and Fitted
## $2,753.00

**The Kitchen.** Directly back of the dining room is the kitchen. It is 8 feet 5 inches by 14 feet 3 inches. It has two large cupboards, a convenience that is appreciated by the busy housewife. Each cupboard is equipped with adjustable shelves. The large cupboard has a working surface that does away with the need of a table. Sink is placed underneath window. There is a space for the range and refrigerator or table. Three windows provide light and cross ventilation. A door leads to rear entry and steps to basement and grade.

**The Bedrooms.** From the dining room a door opens into the hall connecting with bedrooms, bathroom and hall coat closet. Each bedroom has a clothes closet, and is lighted and aired by two windows.

**The Bathroom** is located between bedrooms. It has a medicine cabinet. A window provides plenty of light and air.

**Sleeping Porch.** French doors connect the rear bedroom with sleeping porch. There is a space for two beds and other suitable furnishings. One may sleep out-of-doors and still have all the comforts of an interior room.

**The Basement.** Lighted by cellar sash on three sides. Room for furnace, storage and laundry.

**Height of Ceilings.** Main floor, 9 feet from floor to ceiling. Basement, 7 feet from floor to joists. Basement has cement floor.

### What Our Price Includes

At the price quoted we will furnish all the material to build this five-room and sleeping porch bungalow, consisting of:

Lumber; Lath;
Roofing, Best Grade Clear Red Cedar Shingles;
Siding, Best Grade Clear Red Cedar Shingles;
Framing Lumber, No. 1 Quality Douglas Fir or Pacific Coast Hemlock;
Flooring, Clear Maple for Kitchen and Bathroom, Clear Oak for Balance of Rooms;
Sleeping Porch Floor, Clear Edge Grain Fir;
Porch Ceiling, Clear Douglas Fir or Pacific Coast Hemlock;
Finishing Lumber;
High Grade Millwork (see pages 110 and 111);
Interior Doors, One Panel Design of Douglas Fir;
Trim, Beautiful Grain Douglas Fir or Yellow Pine;
Windows, California Clear White Pine;
Screens for Sleeping Porch;
Medicine Case;
Kitchen Cabinet; Kitchen Cupboard;
Bookcase Colonnade; Bookcases;
Mantel; Wall Safe;
Eaves Trough and Down Spout;
40-Lb. Building Paper; Sash Weights;
Chicago Design Hardware (see page 132);
Paint for Three Coats Outside Trim;
Stain, Two Brush Coats for Shingles on Walls;
Shellac and Varnish for Interior Trim and Doors;
Shellac, Paste Filler and Floor Varnish for Oak and Maple Floors.

Complete Plans and Specifications.

We guarantee enough material to build this house. Price does not include cement, brick or plaster.

See description of "Honor Bilt" Houses on pages 12 and 13.

**Can be built on a lot 42 feet wide**

This house can be built with rooms reversed. See page 3.

SLEEPING PORCH 7'-6"x 11'-2"
FRENCH DOORS
BED ROOM 12'-5"x 9'-11"
BED
DRESSER
CUPBOARD
KITCHEN 8'-5"x 14'-3"
CUPBOARD
SINK
BATH ROOM
CLOSET
CLO.
CLOSET
HALL
DRESSER
BED ROOM 12'-5"x 10'-11"
BED
BUFFET
DINING ROOM
DINING TABLE
15'-7"x 15'-1"
PORCH CEMENT FLOOR
CLO.
SEAT
CLO.
PIANO
LIVING ROOM 23'-11"x 11'-11"
TABLE
DAVENPORT
CHAIR
MANTEL
CHAIR
DESK
44'-0"
26'-0"
PORCH CEMENT FLOOR

**FLOOR PLAN**

### OPTIONS

*Sheet Plaster and Plaster Finish, to take the place of wood lath, $181.00 extra. See page 109.*

*Oriental Asphalt Shingles, guaranteed 17 years, instead of wood shingles, for the roof, $61.00 extra.*

*Oak Doors and Trim, for living room and dining room, $214.00 extra.*

*Storm Doors and Windows, $123.00 extra.*

*Screen Doors and Windows, galvanized wire, $55.00 extra.*

For prices of Plumbing, Heating Wiring, Electric Fixtures and Shades see pages 130 and 131.

*For Our Easy Payment Plan See Page 144*

THE OAKDALE is a bungalow home of unusual charm. It is a masterpiece of one of America's best architects.

The Oakdale's exterior, as well as its floor plan, deserves your careful study. A unique feature is the large front porch, size, 15 feet by 8 feet, with brick foundation and large square columns that support the trusses that carry its roof with its wide overhanging eaves. Note the timber purlins that ornament the gables and add strength to the structure; the three louvre ventilators that adorn the

**Honor Bilt**

## The Oakdale
### No. P3206A "Already Cut" and Fitted
## $1,842.00

gables; the wide overhang of the main roof with the rafter ends exposed; the wide and clear cypress siding and divided light windows and front door. It is these combined features that make the Oakdale one of the most attractive bungalows ever built. All material is of "Honor Bilt" quality —there's none better. The price is exceedingly low, considering the quality and workmanship throughout.

**Porch.** Size, 15 feet wide by 8 feet deep; provides plenty of space for furniture and swing.

**The Living Room.** Through a beautiful glazed door you enter the living room, size, 19 feet 8 inches wide by 11 feet 4 inches deep. To the right is the fireplace and mantel. To the left of entrance is a clothes closet. Wall and floor space provide for piano, davenport, furniture and radio. Six windows provide plenty of light and air.

**The Dining Room.** A wide cased opening divides the living and dining room. Size, 13 feet 8 inches wide by 13 feet 1 inch deep. Here again space favors a satisfactory setting of furniture. A cheerful atmosphere is assured by the double windows.

**The Kitchen.** To the left of dining room a swinging door opens into kitchen. Space for range is near door. Table space, kitchen cabinet, built-in ironing board and broom closet are located to save steps. Below double window is space for sink. Kitchen is well lighted and aired. A door opens to side entry, which has space for refrigerator and stairs to basement and grade.

**The Bedrooms.** A hall is open from the dining room which accommodates bedrooms and bath. Each of the two bedrooms has a clothes closet and two windows. In the bathroom fixtures are on one wall, reducing cost of plumbing. Light and air from one window.

**Basement.** Excavated basement with concrete floor. Room for furnace, laundry and storage.

**Height of Ceilings.** Basement, 7 feet high from floor to joists. Main floor, 9 feet from floor to ceiling.

### What Our Price Includes
At the price quoted we will furnish all the material to build this five-room bungalow consisting of:

**Lumber; Lath;**
**Roofing,** Best Grade Clear Red Cedar Shingles;
**Siding,** Clear Cypress or Clear Red Cedar, Bevel;
**Framing Lumber,** No. 1 Quality Douglas Fir or Pacific Coast Hemlock;
**Flooring,** Clear Douglas Fir or Pacific Coast Hemlock;
**Porch Ceiling,** Clear Douglas Fir or Pacific Coast Hemlock;
**Finishing Lumber;**
**High Grade Millwork** (see pages 110 and 111),
**Interior Doors,** Two Panel Design of Douglas Fir;
**Trim,** Beautiful Grain Douglas Fir or Yellow Pine;
**Kitchen Cabinet;**
**Medicine Case;**
**Mantel;**
**Ironing Board;**
**Eaves Trough and Down Spout;**
**Windows,** California Clear White Pine;
**40-Lb. Building Paper; Sash Weights;**
**Stratford Design Hardware** (see page 132);
**Paint** for Three Coats Outside Trim and Siding;
**Shellac and Varnish** for Interior Trim and Doors.

Complete Plans and Specifications.

Built on concrete foundation with brick above grade. Excavated under entire house.

We guarantee enough material to build this house. Price does not include cement, brick or plaster.

See description of "Honor Bilt" Houses on pages 12 and 13.

#### OPTIONS
*Sheet Plaster and White Plaster Finish, in place of wood lath, $156.00 extra. See page 109.*

*Oak Doors, Floors and Trim for living room and dining room, Maple Floors in kitchen and bathroom, $164.00 extra.*

*Oriental Asphalt Shingles, guaranteed 17 years, instead of wood shingles, $49.00 extra.*

*Storm Doors and Windows, $54.00 extra.*

*Screen Doors and Windows, galvanized wire, $38.00 extra.*

For prices of Plumbing, Heating, Wiring, Electric Fixtures and Shades see pages 130 and 131.

**Can be built on a lot 30 feet wide**

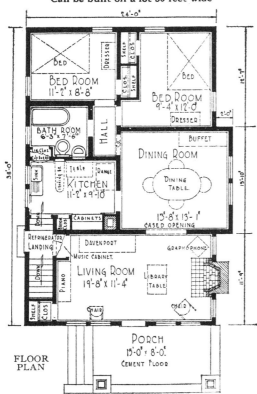

FLOOR PLAN

*For Our Easy Payment Plan See Page 144*

### Honor Bilt

# The Wayne

### No. P13210 "Already Cut" and Fitted
## $2,109.00

THE WAYNE is one of our most popular designs. One looking at the exterior of this house is impressed with its stability. A study of its floor plan reveals the unusual care taken by the architect to give largest rooms possible. Special attention is drawn to the large living room and the large bedroom on second floor and the well arranged stairway.

#### FIRST FLOOR

**The Living Room.** The large porch, size 8x24 feet, extends across the entire front of the house. The Wayne has a beautiful front door in keeping with the high standard material used throughout the house. As you enter you will see that the entire front part of the house is devoted to the large living room. Measures 19 feet 11 inches by 11 feet 10 inches. Directly to the right of the front entrance is a stairway to the second floor. A clothes closet is on the stair landing. Light from three sides makes the room livable and cheerful.

**The Dining Room.** A large cased opening from the living room leads to the dining room, plenty large enough to accommodate dining room furniture. Size, 12 feet 6 inches by 13 feet. Windows on two sides assure desired light and air.

**The Kitchen.** You enter the kitchen from the dining room through a swinging door. It is 10 feet 2 inches by 8 feet 9 inches. Placed to save steps are the sink, space for range, sanitary cupboards. Two windows ventilate and light the kitchen. From here a door connects with side entry which has space for refrigerator. Steps lead to basement and outside door.

#### SECOND FLOOR

**The Bedrooms.** A short hallway connects with the two bedrooms and bath. The bedrooms are well lighted and have large clothes closets.

The Bathroom has one window, and a built-in medicine case.

**Basement.** Under the entire house. Room for furnace, storage and laundry.

**Height of Ceilings.** Basement, 7 feet from floor to joists. First floor, 9 feet from floor to ceiling. Second floor, 8½ feet from floor to ceiling.

### What Our Price Includes

At the price quoted we will furnish all the material to build this five-room two-story house, consisting of:

**Lumber; Lath;**

**Roofing,** Oriental Slate Surfaced Shingles, Guaranteed 17 Years;

**Siding,** Clear Cypress or Clear Red Cedar, Bevel;

**Framing Lumber,** No. 1 Quality Douglas Fir or Pacific Coast Hemlock;

**Flooring,** Clear Douglas Fir or Pacific Coast Hemlock;

**Porch Flooring,** Clear Edge Grain Fir;

**Porch Ceiling,** Clear Grade Douglas Fir or Pacific Coast Hemlock;

**Finishing Lumber;**

**High Grade Millwork** (see pages 110 and 111);

**Interior Doors,** Two Panel Design of Douglas Fir;

**Trim,** Beautiful Grain Douglas Fir or Yellow Pine;

**Colonial Shutters;**

**Windows,** California Clear White Pine;

**Medicine Case;**

**Kitchen Cupboards;**

**40-Lb. Building Paper; Sash Weights;**

**Eaves Trough and Down Spout;**

**Stratford Design Hardware** (see page 132);

**Paint** for Three Coats Outside Trim and Siding;

**Shellac and Varnish** for Interior Trim and Doors.

Complete Plans and Specifications.

Built on a concrete foundation. Brick above grade.

We guarantee enough material to build this house. Price does not include cement, brick or plaster.

See Description of "Honor Bilt" Houses on pages 12 and 13.

#### OPTIONS

*Sheet Plaster and Plaster Finish, to take the place of wood lath and plaster, $191.00 extra. See page 109.*

*Oak Doors, Trim and Floor, for living room and dining room, also Oak Stairs. Maple Floors, in kitchen and bathroom, $193.00 extra.*

*Storm Doors and Windows, $68.00 extra.*

*Screen Doors and Windows, galvanized wire, $43.00 extra.*

For Prices of Plumbing, Heating, Wiring, Electric Fixtures and Shades see pages 130 and 131.

Can be built on a lot 30 feet wide

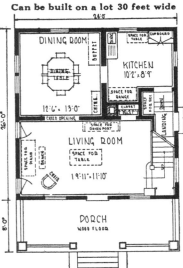

FIRST FLOOR PLAN

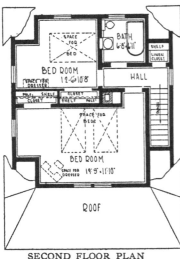

SECOND FLOOR PLAN

*For Our Easy Payment Plan See Page 144*

# The WAYNE~INTERIORS

*ABOVE*—The entire front part of the first floor is given over to the living room. Here stairs ascend to the second floor, a mirror door opens to coat closet, and a wide cased opening leads to the dining room.

*CENTER LEFT*—The kitchen is equipped with a built-in cupboard, and has space for the sink, range, table and chair. Space for the refrigerator is provided in the side entry, close to the kitchen door.

*BELOW*—The master's bedroom provides space for twin beds, table and night lamp, dresser, chiffonier, rocker, etc. It has a sizable clothes closet with two doors. Plenty of light and cross ventilation is assured by a double front window and one side window. Floor dimensions are: 19 feet 3 inches by 11 feet 10 inches.

*CENTER RIGHT*—The dining room, 12 feet 6 inches by 13 feet, is just the right size for a complete dining set, and accommodation of family and guests.

The Wayne Home may be attractively furnished in any number of styles, one of which is illustrated above.

**Honor Bilt**

## The *Barrington*

### No. P3241 "Already Cut" and Fitted

### $2,606<u>00</u>

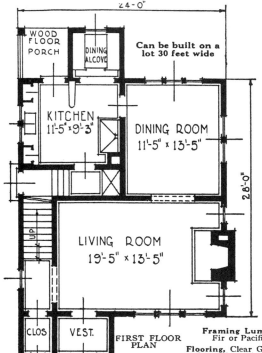

Can be built on a lot 30 feet wide

WOOD FLOOR PORCH

DINING ALCOVE

KITCHEN 11'-5" x 9'-3"

DINING ROOM 11'-5" x 13'-5"

LIVING ROOM 19'-5" x 13'-5"

CLOS. VEST.

FIRST FLOOR PLAN

24'-0"

28'-0"

*For Our Easy Payment Plan See Page 144*

THE BARRINGTON retains the dignity of an old English home and has the practical interior of modern American architecture. Whether you consider economy, beauty or convenience as of first importance, The Barrington home assuredly meets these and every point of merit with satisfaction. Exterior features at once stamp the mark of quality. The well balanced projection at the front forms the entrance, leading to it is a tapestry brick terrace, guarded by a decorative iron railing. Sided with wide shingles and exposed fireplace chimney.

### FIRST FLOOR

**The Entrance.** From the open terrace you enter the vestibule.

**The Living Room** space is 19 feet 5 inches by 13 feet 5 inches, and for all practical purposes, extends the full width of the house, because the stair hall is really a part of the living room, divided only by a cased opening. A mantel and fireplace is on the right wall, with a high casement sash on each side. Lighted by a triple window in the front. A coat closet is in the stair hall. Here the stairs ascend to the second floor.

**The Dining Room.** 11 feet 5 inches by 13 feet 5 inches. The rear wall is planned for buffet space, over which is a triple high casement sash. A double window on the side admits additional light and ventilation.

**The Kitchen** is 11 feet 5 inches by 9 feet 3 inches, just the wanted size, reducing steps and labor. One of the special built-in features is our De Luxe Outfit. Another practical item is the Built-In and Disappearing Ironing Board. A double window assures plenty of light and ventilation. A rear door leads to the rear porch. Another door leads to the side entry, which has space for a refrigerator, stairs to basement and grade.

**The Dining Alcove** is entered from the kitchen through a wide cased opening. It has our Built-In Breakfast Set.

### SECOND FLOOR

**The Bedrooms.** Stairway from the first floor leads to the hall which connects with three bedrooms, bathroom and linen closet. The bedroom at the right front is planned to be used as the master's room, accommodating twin beds. It has a clothes closet, a double window at the front and one window on the side. The left front bedroom contains door to attic stairs, a double window at the side and a good closet in the front. The rear bedroom has one window on each wall and a clothes closet. The bathroom has a built-in medicine case and a window.

**The Basement.** Room for heating plant, laundry and storage.

**Height of Ceilings.** First floor, 8 feet 6 inches from floor to ceiling. Second floor, 8 feet 6 inches from floor to ceiling. Basement, 7 feet from floor to joists.

### *What Our Price Includes*

At the price quoted we will furnish all the material to build this six-room house, consisting of:

Lumber; Lath;

Roofing, Best Grade Clear Red Cedar Shingles;

Siding, Best Grade Clear Red Cedar Shingles;

Framing Lumber, No. 1 Quality Douglas Fir or Pacific Coast Hemlock;

Flooring, Clear Grade Douglas Fir or Pacific Coast Hemlock;

Porch Flooring, Clear Grade Douglas Fir or Pacific Coast Hemlock;

Porch Ceiling, Clear Grade Douglas Fir or Pacific Coast Hemlock;

Finishing Lumber;

High Grade Millwork (see pages 110 and 111);

Interior Doors, Two-Panel Design of Douglas Fir;

Trim, Beautiful Grain Douglas Fir or Yellow Pine;

Windows of California Clear White Pine;

Medicine Case; Flower Box;

Built-In Ironing Board; De Luxe Kitchen Outfit;

Mantel; Breakfast Alcove, Table and Seats;

Colonial Shutters;

40-Lb. Building Paper; Sash Weights;

Chicago Design Hardware (see page 132);

Paint for Three Coats Outside Trim;

Stain for Shingles on Walls for Two Brush Coats;

Varnish and Shellac for Interior Doors and Trim.

We guarantee enough material to build this house. Price does not include cement, brick or plaster.

See description of "Honor-Bilt" Houses on pages 12 and 13 of Modern Homes Catalog.

BATH 6'-2" x 9'-5"

LIN CLOS

BED

BED ROOM 13'-2" x 9'-7"

DOWN

HALL

CLOS CLOS

ATTIC

BED ROOM 11'-3" x 13'-5"

BED

BED ROOM 11'-7" x 13'-5"

BED

CLOSET

SECOND FLOOR PLAN

### OPTIONS

*Sheet Plaster and Plaster Finish,* to take the place of wood lath, $217.00 extra. See page 109.

*Storm Doors and Windows,* $126.00 extra.

*Screen Doors and Windows,* galvanized wire, $87.00 extra.

*Oriental Slate Surfaced Shingles,* in place of wood shingles for roof, $33.00 extra.

For prices of Plumbing, Heating, Wiring, Electric Fixtures and Shades see pages 130 and 131.

**Honor Bilt**

## The Columbine
### No. P8013 "Already Cut" and Fitted
## $2,162.00

THE COLUMBINE, a unique creation in American architecture, has always received praise wherever it has been built. The symmetrical lines of the front cannot be visualized by the illustration. However, the front porch, size 24 feet by 9 feet 6 inches, is roofed in the center, with pergolas at either side which shade the porch, yet give the full benefit of all the air and light. The porch roof and pergolas are supported by six colonial columns. The dentals in the porch gables give it the final touch of elegance and good taste. Don't overlook the triple windows on either side of the porch, the massive brick chimney on the left and the special divided lights in the upper sash of all the windows. You will agree that it is a creation hard to beat.

The floor plan meets the desires of those who want the living and dining rooms in the front, and the sleeping rooms in the rear.

**The Living Room** is 13 feet 11 inches by 15 feet 5 inches, being of good proportion, with a large amount of wall space, there is plenty of room for piano and other furniture. There is a fireplace on the outside wall. This room has five windows, and a fine glazed door, all with divided lights.

**The Dining Room.** From the living room a pair of French doors connect with the dining room. The French door opening lends depth to both rooms, which makes them available as one large room when entertaining.

**The Kitchen** has been given special study. Note that the range sets in an alcove between the chimney and closet, the latter seldom being found in kitchens nowadays. The cupboard and sink are on the opposite side of the room, close to the double window, and very handy to the dining room. The rear kitchen door leads to a landing from which you enter the basement, or go outside. On the landing is space for a refrigerator, and directly above are shelves that will save many a step to the basement. A glazed door gives light to the stairway.

**The Bedrooms.** Being all in one part of the house and close to the kitchen, the bedroom work can be attended to while the kitchen work and cooking are in progress. A hall connects the sleeping chambers and bathroom, which is in the rear. All bedrooms have closets with shelves and wardrobe poles.

**Basement.** Room for furnace, laundry and storage.

**Height of Ceilings.** Main floor, 9 feet from floor to ceiling. Basement, 7 feet from floor to joists.

### What Our Price Includes

At the price quoted we will furnish all the material to build this six-room house, consisting of:
Lumber; Lath;
**Roofing,** Best Grade Clear Red Cedar Shingles;
**Siding,** Clear Cypress or Clear Red Cedar, Bevel;
**Framing Lumber,** No. 1 Quality Douglas Fir or Pacific Coast Hemlock;
**Flooring,** Clear Douglas Fir or Pacific Coast Hemlock;
**Porch Ceiling,** Clear Douglas Fir or Pacific Coast Hemlock;
**Finishing Lumber;**
**High Grade Millwork** (see pages 110 and 111);
**Interior Doors,** Five Cross Panel Design of Douglas Fir;
**Trim,** Beautiful Grain Douglas Fir or Yellow Pine;
**Kitchen Cupboard; Medicine Case;**
**Windows,** California Clear White Pine;
**Building Paper; Sash Weights;**
**Eaves Trough and Down Spouts;**
**Stratford Design Hardware** (see page 132);
**Paint** for Three Coats Outside Trim and Siding;
**Shellac and Varnish** for Interior Trim and Doors.

Complete Plans and Specifications.

Built on a concrete and brick foundation and excavated under entire house.

We guarantee enough material to build this house. Price does not include cement, brick or plaster.

See description of "Honor Bilt" Houses on pages 12 and 13.

### ATTIC FLOOR PLAN
See Options.

**ROOM** 11'0" x 10'9"

CLOSET SHELF
SHELF
CLOSET

HALL DOWN

**ROOM** 11'0" x 18'0"

SHAPE OF ATTIC

**Can be built on a lot 34 feet wide**

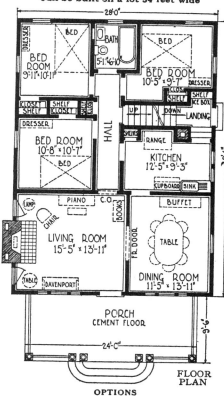

FLOOR PLAN

### OPTIONS

*Furnished With Two Rooms in Attic, with single floor, $244.00 extra. See floor plan.*

*Sheet Plaster and Plaster Finish, to take the place of wood lath, $184.00 extra; with attic, $254.00. See page 109.*

*Oriental Asphalt Shingles, guaranteed 17 years, instead of wood shingles, $55.00 extra.*

*Oak Doors, Trim and Floors in living and dining room, Maple Floors in kitchen and bathroom, $160.00 extra.*

*Storm Doors and Windows, $61.00 extra; with attic, $76.00 extra.*

*Screen Doors and Windows, galvanized wire, $41.00 extra; with attic, $53.00 extra.*

For prices of Plumbing, Heating, Wiring, Electric Fixtures and Shades see pages 130 and 131.

*For Our Easy Payment Plan See Page 144*

## The Hampton

### No. P3208 "Already Cut" and Fitted

### $1,681.00

**B**UNGALOW architecture features The Hampton. The interior is designed along practical lines. Full use of space affords a greater amount of room than is usual in a house of this size. The location of each room and its relation to the rest of the house has been planned to promote the comfort of the family.

Our best grade of material and "Honor Bilt" construction assure satisfaction. Priced very low, and when built it will make a comfortable home or can be sold at a handsome profit.

**The Living Room.** At front of the house is a good sized porch which can be enclosed with wire screen at any time. From the porch one enters the living room, which measures 13 feet 5 inches by 11 feet 11 inches.

**The Dining Room.** A large cased opening between living room and dining room makes both rooms serve as one. Well arranged windows give the necessary light and ventilation and provide suitable wall spaces in both rooms for the placing of furniture.

**The Kitchen.** A swinging door connects the dining room and kitchen. Here efficient arrangement of the cupboard and sink with space for range and table saves time and labor. A double window fills the kitchen with brightness, making the daily work a pleasure.

From the kitchen a door opens into the enclosed rear entry. Space for ice box is near the door. From here steps reach the basement and the outside door.

**The Bedrooms.** The front bedroom opens from the living room. It has a clothes closet, and two windows. Another bedroom is entered from the dining room. Here, too, is a clothes closet and a window.

A small hallway is open from the dining room. This gives privacy to the rear bedroom and bath. The bedroom is lighted by two windows. All the bathroom plumbing is on one wall.

**Basement.** Room for laundry, furnace and storage.

**Height of Ceilings.** Basement is 7 feet from floor to joists. Main floor rooms are 9 feet from floor to ceiling.

### What Our Price Includes

At the price quoted we will furnish all the material to build this six-room bungalow consisting of:

**Lumber; Lath;**
**Roofing,** Best Grade Clear Red Cedar Shingles;
**Siding,** Clear Cypress or Clear Red Cedar, Bevel;
**Framing Lumber,** No. 1 Quality Douglas Fir, or Pacific Coast Hemlock;
**Flooring,** Clear Grade Douglas Fir or Pacific Coast Hemlock for All Rooms, Clear Edge Grain Fir for Porch;
**Porch Ceiling,** Clear Grade Douglas Fir or Pacific Coast Hemlock;
**Finishing Lumber;**
**High Grade Millwork** (see pages 110 and 111);
**Interior Doors,** Two-Panel Designs of Douglas Fir;
**Trim,** Beautiful Grain Douglas Fir or Yellow Pine;
**Kitchen Cabinet; Medicine Case;**
**Windows,** of California Clear White Pine;
**40-Pound Building Paper; Sash Weights;**
**Eaves Trough and Down Spout;**
**Stratford Design Hardware** (see page 132);
**Paint** for Three Coats Outside Trim and Siding;
**Shellac and Varnish** for Interior Trim and Doors.

Complete Plans and Specifications.

Built on a concrete and brick foundation and excavated under entire house.

We guarantee enough material to build this bungalow.

Price does not include cement, brick or plaster.

See description of "Honor Bilt" Houses on pages 12 and 13.

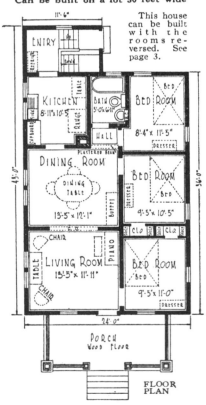

**Can be built on a lot 30 feet wide**

This house can be built with the rooms reversed. See page 3.

**FLOOR PLAN**

### OPTIONS

*Sheet Plaster and Plaster Finish to take the place of wood lath and plaster, $164.00 extra. See page 109.*

*Storm Doors and Windows, $55.00 extra.*

*Screen Doors and Windows, galvanized wire, $36.00 extra.*

*Oriental Slate Surfaced Shingles, instead of wood shingles, $44.00 extra.*

For prices of Plumbing, Heating, Wiring, Electric Fixtures and Shades see pages 130 and 131.

**A View of the Kitchen**

### For Our Easy Payment Plan See Page 144

**Honor Bilt**

## The Hathaway

### No. P3195 "Already Cut" and Fitted
### $1,805.00

AMERICAN tourists in Europe are always favorably impressed by the cottage homes of England. They speak enthusiastically of the appearance of solid comfort they convey, and describe them as covered with vines and climbing roses. Most of these English homes are constructed of stone, brick or concrete.

The Hathaway is a striking example of this style of architecture, in frame construction with wood shingle siding. It makes a home suitable for suburb or country. The treatment is unique and artistic, the result being achieved at comparatively small cost. The ornamental trellised porch is a typical English feature. It is cozy and graceful and the colonial and French windows with their flower boxes make this type of house at home in any American community. The proud possessor will quickly avail himself of the special advantages which an artistic grouping of shrubs and flowers will secure. Vines or ivy on the trellis and side walls will transform this house into a bower of beauty.

### FIRST FLOOR

**The Living Room** is entered from the front porch. Size, 13 feet 8 inches by 12 feet 1 inch. To the right an open stairway, over which is a high sash, leads to the second floor. Directly opposite is wall space for a piano, and ample floor space for a davenport, radio, table, chairs and lamp.

**The Dining Room** and the living room are united by a wide cased opening. Size, 12 feet 1 inch by 12 feet 7 inches. Cross ventilation and a pleasant atmosphere are insured by two windows on each outer wall.

**The Kitchen** connects with the dining room, the pantry, the broom closet, and the covered rear porch which has stairs to grade. Size, 10 feet 9 inches by 11 feet 8 inches. Plenty of space for sink, range, table and chair. The pantry has shelves, and space for the refrigerator.

### SECOND FLOOR

**The Bedrooms**, the bathroom and a closet suitable for linens, etc., connect with the hall off the stairway. Each of the three bedrooms has a double window, a closet for clothing, and liberal space for the usual furniture.

**The Basement.** Space for furnace, laundry and storage.

**Height of Ceilings.** First floor, 8 feet 2 inches from floor to ceiling. Second floor, 8 feet 2 inches from floor to ceiling. Basement, 7 feet from floor to joists.

### What Our Price Includes

At the price quoted we will furnish all the material to build this six-room, two-story house, consisting of:

**Lumber; Lath;**
**Roofing,** Best Grade Clear Red Cedar Shingles;
**Siding,** Best Grade Clear Red Cedar Shingles;
**Framing Lumber,** No. 1 Quality Douglas Fir or Pacific Coast Hemlock;
**Flooring,** Clear Douglas Fir or Pacific Coast Hemlock;
**Porch Ceiling,** Clear Douglas Fir or Pacific Coast Hemlock;
**Finishing Lumber;**
**High Grade Millwork** (see pages 110 and 111);
**Interior Doors,** Five Cross Panel Design of Douglas Fir;
**Trim,** Beautiful Grain Douglas Fir or Yellow Pine;
**Windows,** California Clear White Pine;
**Medicine Case;**
**Eaves Trough and Down Spouts;**
**40-Lb. Building Paper; Sash Weights;**
**Stratford Design Hardware** (see page 132);
**Paint** for Three Coats Outside Trim;
**Stain,** Two Brush Coats for Shingles on Walls;
**Shellac and Varnish** for Interior Trim and Doors.

Complete Plans and Specifications.

We guarantee enough material to build this house. Price does not include cement, brick or plaster.

"Honor Bilt" Construction explained on pages 12 and 13.

### OPTIONS

*Sheet Plaster and Plaster Finish to take the place of wood lath, $166.00 extra. See page 109.*

*Oriental Asphalt Shingles, guaranteed 17 years, instead of wood shingles for roof, $33.00 extra.*

*Storm Doors and Windows, $79.00 extra.*

*Screen Doors and Windows, galvanized wire, $51.00 extra.*

For Prices of Plumbing, Heating, Wiring, Electric Fixtures and Shades see pages 130 and 131.

### For Our Easy Payment Plan
### See Page 144

Can be built on a lot 36 feet wide

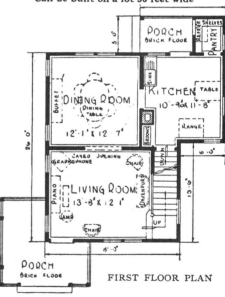

FIRST FLOOR PLAN

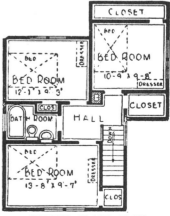

SECOND FLOOR PLAN

THE VAN DORN Home is a fine example of modern Dutch colonial architecture. While its exterior reflects the classic of our historical colonial period, its interior has every advantage of our present day superior arrangement. The fact that this home is of "Honor Bilt" material and ready-cut construction is a positive assurance of real value and economy of upkeep.

The Van Dorn home is at once the pride of its owner and, if for sale, is sure to bring a handsome profit, as this style is always popular.

You approach The Van Dorn aware that here is a beautiful home. It is adorned with a true colonial entrance and windows. Green shutters contrast the pure white siding, while below is the red brick foundation and buttress.

The interior readily meets with approval. Every room is laid out in relation to the others, and spaced according to use. There is not an inch of wasted space. The rooms are unusually spacious for a home of this size, 26 feet by 24 feet. Here, again, "Honor Bilt" architecture achieves success. Study the floor plans.

### FIRST FLOOR

**The Vestibule.** The front door opens into the vestibule. A stairway leads to the second floor, a guest coat closet is located at the left of stairway, and a wide cased opening welcomes you into the living room.

**The Living Room.** There is more space here than is customary in a home of this size. It measures 18 feet 4 inches by 11 feet 5 inches. Good wall space makes for a splendidly furnished room, including a piano. Two front windows and a double side window provide all the wanted light and ventilation.

**The Dining Room.** A wide cased opening unites the living room and dining room, thus making it convenient for the hostess to entertain many guests. Size of dining room, 12 feet 5 inches by 11 feet 5 inches. An air of cheerfulness is assured by the flood of light from the windows.

**The Kitchen.** You enter the kitchen from the dining room through a swinging door. Size of kitchen, 12 feet 5 inches by 9 feet 2 inches. The space for a built-in cupboard, the range, table and sink make possible a very satisfactory arrangement. Hundreds of steps and precious time will be saved every day. A window on each outer wall provides light and cross current of air.

A door to the side entry leads to space for ice box, stairs to basement and door to grade.

### The Van Dorn
#### No. P3234 "Already Cut" and Fitted
### $2,249.00

#### SECOND FLOOR

**The Bedrooms.** The stairs from the vestibule bring you to the hall on the second floor. Each of the three bedrooms has a clothes closet, and two windows with cross current of air.

**The Bathroom** is located close to every bedroom and stairway. It has a linen closet and a built-in medicine case.

**The Basement.** Room for laundry, furnace and storage. Lighted by sash.

**Height of Ceilings.** First floor, 8 feet 6 inches from floor to ceiling. Second floor, 8 feet 6 inches from floor to ceiling. Basement, 7 feet from floor to joists.

### What Our Price Includes

At the price quoted we will furnish all the material to build this six-room two-story house consisting of:

**Lumber; Lath;**
**Roofing,** Best Grade Clear Red Cedar Shingles;
**Siding,** Clear Grade Cypress, or Clear Red Cedar, Bevel;
**Framing Lumber,** No. 1 Quality Douglas Fir or Pacific Coast Hemlock;
**Flooring—First Floor,** Clear Oak for Vestibule, Living Room and Dining Room, Clear Maple for Kitchen and Entry; **Second Floor,** Clear Maple for Bathroom, Clear Douglas Fir or Pacific Coast Hemlock for Bedrooms and Hall;
**Porch Ceiling,** Clear Douglas Fir or Pacific Coast Hemlock;
**Finishing Lumber;**
**High Grade Millwork** (see pages 110 and 111);
**Interior Doors,** Two Vertical Panel Design of Douglas Fir; also Mirror Door;
**Trim,** Beautiful Grain Douglas Fir or Yellow Pine;
**Windows,** California Clear White Pine;
**Medicine Case;**
**Colonial Shutters;**
**Eaves Trough and Down Spout;**
**40-Lb. Building Paper; Sash Weights;**
**Chicago Design Hardware** (see page 132);
**Paint** for Three Coats Outside Trim and Siding;
**Mahogany Stain** for All Doors; also Stair Treads, Stair Rails and Newel. All other interior trim and woodwork White Enamel.
Complete Plans and Specifications.
We guarantee enough material to build this house. Price does not include cement, brick or plaster. See description of "Honor Bilt" Houses on pages 12 and 13.

#### FIRST FLOOR PLAN

**Can be built on a lot 32 feet wide.**

#### SECOND FLOOR PLAN
#### OPTIONS

*Sheet Plaster and Plaster Finish, to take the place of wood lath, $207.00 extra. See page 109.*
*Oriental Slate Surfaced Shingles, instead of wood shingles, $31.00 extra.*
*Storm Doors and Windows, $59.00 extra.*
*Screen Doors and Windows, galvanized wire, $43.00 extra.*
For prices of Plumbing, Heating, Wiring, Electric Fixtures and Shades see pages 130 and 131.

### For Our Easy Payment Plan See Page 144

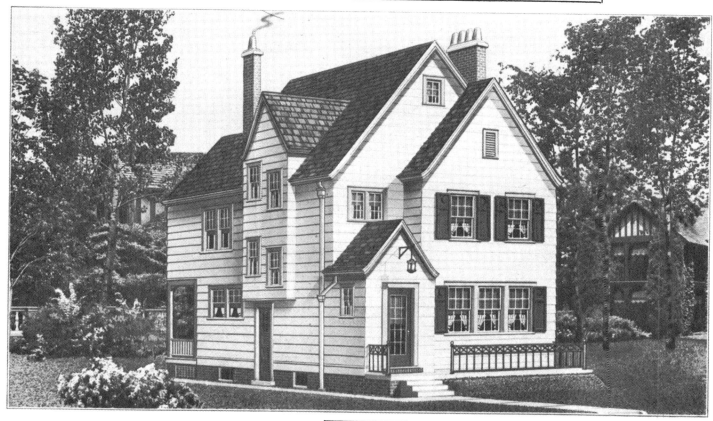

THE CHESTERFIELD HOME has an English ancestry which has stood the test of public favor for many centuries. This particular and interesting design is especially admired for its attractive appearance and splendid interior arrangement. For instance: Consider the informal massing of the walls, the closely clipped gables and the low swung sloping roof. Then, again, additional character is gained by the green shutters, red brick chimneys and the white siding contrasted by gray trim. The concrete terrace with its decorative iron railing makes another appreciative decoration.

Here is a home that is just as comfortable and attractive within as it is lovely outside. Made of high grade "Honor Bilt" material, and priced far below actual value. Our factory-to-you price saves you the middleman's profit.

### Honor Bilt
## The Chesterfield
#### No. P3235 "Already Cut" and Fitted
## $2,934.00

#### FIRST FLOOR

**The Living Room.** You enter through a vestibule, which is lighted by a high window. A coat closet is at one side.

The living room fronts the entire width of the house, and is well proportioned. Size, 23 feet 3 inches by 14 feet 5 inches. Facing you is the open stairway that rises to the second floor. Attractively set in the center space of the right wall is a colonial mantel and fireplace, with a high casement sash at each side. Three large windows at the front provide additional light.

**The Dining Room.** A pair of French doors connect the living room and dining room. Here good spacing, 12 feet 5 inches by 13 feet 5 inches, will accommodate a complete dining suite. Light and ventilation from a double window at the side and a long high casement sash at the rear.

**The Kitchen** is entered from the dining room through a swinging door. The kitchen measures 10 feet 5 inches by 9 feet 8 inches—room enough for all needs, yet compact to save time and labor. Here is just a glimpse of this prize room: Kitchen De Luxe Outfit (illustrated on page 111), a provision door opens from the rear porch, built-in cupboards, space for range.

The side entry accommodates space for refrigerator and has stairway to basement and door to grade.

**The Breakfast Room** is right off the kitchen.

#### SECOND FLOOR

**The Bedrooms.** The stairway from the living room leads directly to the hall on this floor. Here the hall connects with each of the three spacious bedrooms, bathroom, linen closet and stairway to attic. Each of the bedrooms is large enough for twin beds and the usual furniture. Each of the bedrooms has a large clothes closet and at least three windows. The bathroom has a built-in medicine case and a double casement sash.

**The Basement.** Basement under entire house and is lighted by sash. Room for furnace, laundry and storage.

**Height of Ceilings.** First floor, 8 feet 6 inches from floor to ceiling. Second floor, 8 feet 6 inches from floor to ceiling. Basement, 7 feet from floor to joists.

### What Our Price Includes

At the price quoted we will furnish all the material to build this six-room, two-story house, consisting of:
Lumber; Lath;
Roofing, Best Grade Clear Red Cedar Shingles;
Siding, Clear Cypress, or Red Cedar, Bevel;
Framing Lumber, No. 1 quality Douglas Fir or Pacific Coast Hemlock;
Flooring, Clear Grade Douglas Fir or Pacific Coast Hemlock;
Finishing Lumber;
High Grade Millwork (see pages 110 and 111);
Interior Doors, Two-Panel Design of Douglas Fir;
Trim, Beautiful Grain Douglas Fir or Yellow Pine;
Kitchen Cupboards; Medicine Case;
Wardrobe; Coal Chute;
Breakfast Alcove;

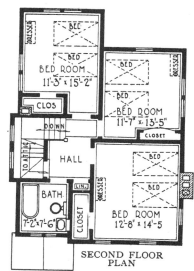

SECOND FLOOR PLAN

Brick Mantel;
Kitchen De Luxe Sink Outfit;
Colonial Shutters;
Windows, California Clear White Pine;
40-Lb. Building Paper; Sash Weights;
Eaves Trough and Down Spout;
Stratford Design Hardware (see page 132);
Paint for Three Coats Outside Trim and Siding;
Paint, Stain for Doors, Stair Treads, Stair Rail and Newel; Enamel for all trim and other woodwork.
Complete Plans and Specifications.

We guarantee enough material to build this house. Price does not include cement, brick or plaster.

See description of "Honor Bilt" Houses on pages 12 and 13.

#### OPTIONS

*Sheet Plaster and Plaster Finish, to take the place of wood lath, $213.00 extra. See page 109.*

*Oriental Asphalt Shingles, instead of wood shingles, $35.00 extra.*

*Storm Sash and Windows, $81.00 extra.*

*Screen Doors and Windows, galvanized wire, $61.00 extra.*

*Ornamental Wrought Iron Fence as shown in illustration, $125.00 extra.*

For prices of Plumbing, Heating, Wiring, Electric Fixtures and Shades see pages 130 and 131.

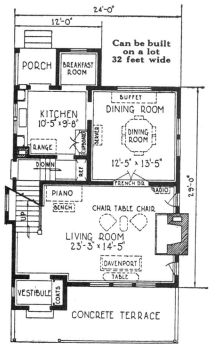

FIRST FLOOR PLAN

*For Our Easy Payment Plan See Page 144*

# NINE ROOMS AND BATH

**Can be built on a lot 45 feet wide**

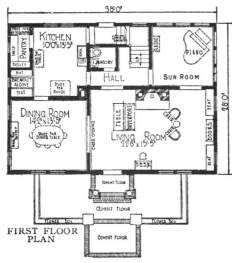

**FIRST FLOOR PLAN**

**THE BREAKFAST ALCOVE**

THE HONOR is a home that not only looks well at a distance, but makes a still more favorable impression upon closer investigation. You will recognize in this house some of the features that have made the historical colonial homes of America admired for years, together with many modern touches that add to its attractiveness. Notice the thatched effect on the roof, the inviting front entrance, the big handsome windows and the decorative trellis. As you step closer, you will admire the cement floored porch, surrounded by flower boxes for perennial plants and evergreens.

### FIRST FLOOR

Passing up the steps to the cement porch you will find yourself in front of the graceful and massive entrance, through which you reach the large living room. To the left a cased opening leads to the dining room, and right in front of you is an inviting brick mantel. To the right in the living room a generous space for the piano is provided between two windows, which have convenient window seats. In the dining room there is ample space for a large buffet between the two windows. From this room a door leads to the kitchen, which is admirably arranged and has a pantry and the very popular breakfast alcove. One kitchen door leads to the hall, off of which is a lavatory, and stairs to the second floor; another kitchen door leads to rear entrance and basement. Directly to the right of the hall is a good size sun room which can also be reached from the living room in the front.

### SECOND FLOOR

On this floor there are four large bedrooms, all provided with closets having shelves, as well as a sleeping porch and bathroom. The ventilating and lighting arrangement is perfect. There is plenty of room for large beds and dressers, as well as other usual furniture.

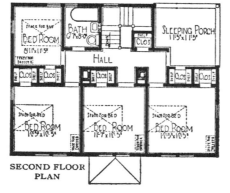

**SECOND FLOOR PLAN**

## The Honor

**No. P13071 "Already Cut" and Fitted**

# $3,278.00

**The Basement.** Excavated under the entire house. Lighted by sash. Room for heating plant, storage and laundry.

**Height of Ceilings.** First floor, 9 feet from floor to ceiling. Second floor, 8 feet 2¼ inches from floor to ceiling. Basement, 7 feet from floor to joists.

### What Our Price Includes

At the price quoted we will furnish all the material to build this nine-room house, consisting of:
Lumber; Lath;
**Roofing,** Oriental Slate Surfaced Shingles;
**Siding,** Best Grade Clear Red Cedar Shingles;
**Framing Lumber,** No. 1 Quality Douglas Fir or Pacific Coast Hemlock;
**Flooring,** Clear Oak for Living Room, Dining Room, Sunroom and Hall. Clear Maple for Kitchen and Bathroom. Clear Douglas Fir or Pacific Coast Hemlock for Balance of Rooms;
**Porch Ceiling,** Clear Douglas Fir or Pacific Coast Hemlock;
Finishing Lumber;
**High Grade Millwork** (see pages 110 and 111);
**Interior Doors,** Two-Panel Design for First Floor and Two-Panel Design for Second Floor, All of Douglas Fir;
**Trim,** Beautiful Grain Douglas Fir or Yellow Pine;
**Windows,** California Clear White Pine;
**Medicine Case;**
**Built-In Seats;**
**Mantel, Brick;**
**Breakfast Alcove and Seats;**
**Screens for Sleeping Porch;**
**Eaves Trough and Down Spout;**
**40-Lb. Building Paper; Sash Weights;**
**Chicago Design Hardware** (see page 132);
**Paint** for Three Coats Outside Trim;
**Stain** for Two Brush Coats for Shingles on Walls;
**Shellac and Varnish** for Interior Trim and Doors;
**Shellac, Paste Filler and Floor Varnish** for Oak and Maple Floors.

Complete Plans and Specifications.

We guarantee enough material to build this house. Price does not include cement, brick or plaster.

See description of "Honor Bilt" Houses on pages 12 and 13.

### OPTIONS

*Sheet Plaster and Plaster Finish, to take the place of wood lath, $306.00 extra. See page 109.*
*Storm Doors and Windows, $116.00 extra.*
*Screen Doors and Windows, galvanized wire, $78.00 extra.*
*Oak Doors and Trim for living and dining rooms, hall and stairs, $222.00 extra.*

For prices of Plumbing, Heating, Wiring, Electric Fixtures and Shades see pages 130 and 131.

*For Our Easy Payment Plan See Page 144*

THE GARFIELD two-story twin apartment home, is considered a splendid investment opportunity. It can be sold at a handsome profit when built in a good locality. Sometimes the owner lives in one apartment and rents out the other to a tenant. Such rental usually pays the cost of the property.

The Garfield has a pleasing exterior and a modern interior, with no waste of space. "Honor Bilt" material and construction are additional features that assure the utmost in value.

### FIRST FLOOR

**Private Entrances.** The first floor apartment as well as the second floor apartment has its own front entrance on the spacious front porch. Size of porch, 26 feet by 8 feet.

**Can be built on a lot 32 feet wide**

FIRST FLOOR PLAN

**Honor Bilt**

## The Garfield

*No. P3232 "Already Cut" and Fitted*

### $2,745.00

**The Living Room.** Size, 11 feet 2 inches by 12 feet 8 inches. Wall space for furniture and piano. Light and air from one front window and one side window.

**The Dining Room.** 10 feet 7 inches by 12 feet 8 inches. A wide cased opening connects the living room with the dining room. Has double window at the side.

**The Kitchen.** From the dining room a swinging door opens into the kitchen. Size of kitchen, 8 feet 8 inches by 10 feet 10 inches. The kitchen has a built-in cupboard and space for sink, range and table. Two windows.

**Rear Porch.** A kitchen door leads to the rear porch which has stairs to basement, and grade.

**The Bedrooms.** Right off the dining room is a hall that connects with the two bedrooms and bathroom. Each bedroom has a clothes closet. The front bedroom has one window, and the rear bedroom has two windows.

**The Bathroom** has a built-in medicine case.

### SECOND FLOOR

Stairs lead up from the front porch to the hall on the second floor. All rooms are planned the same as those of the first floor, except the front bedroom, which is larger and has three windows.

**Read first floor description for details.**

### BASEMENT

Room for a furnace for both apartments or a furnace for each apartment; also, laundry and storage.

**Height of Ceilings.** First floor, 8 feet 6 inches from floor to ceiling. Second floor, 8 feet 6 inches from floor to ceiling. Basement, 7 feet from floor to joists.

### What Our Price Includes

At the price quoted we will furnish all the material to build this ten-room house, consisting of:
Lumber; Lath;
**Roofing,** Best Grade Clear Red Cedar Shingles;
**Siding,** Clear Cypress or Clear Red Cedar, Bevel;
**Framing Lumber,** No. 1 Quality Douglas Fir or Pacific Coast Hemlock;
**Flooring,** Clear Grade Douglas Fir or Pacific Coast Hemlock;
**Porch Flooring,** Clear Edge Grain Fir;
**Porch Ceiling,** Clear Douglas Fir or Pacific Coast Hemlock;
**Finishing Lumber;**
**High Grade Millwork** (see pages 110 and 111);
**Interior Doors,** Two-Panel Design of Douglas Fir;

SECOND FLOOR PLAN

**Trim,** Beautiful Grain Douglas Fir or Yellow Pine;
**Windows,** California Clear White Pine;
**Medicine Case;**
**Kitchen Cupboard;**
**Eaves Trough and Down Spout;**
**40-Lb. Building Paper; Sash Weights;**
**Chicago Design Hardware** (see page 132);
**Paint** for Three Coats Outside Trim and Siding;
**Shellac and Varnish** for Interior Trim and Doors.
Complete Plans and Specifications.

We guarantee enough material to build this house. Price does not include cement, brick or plaster. See description of "Honor Bilt" Houses on pages 12 and 13.

### OPTIONS

*Sheet Plaster and Plaster Finish to take the place of wood lath and plaster, $286.00 extra, see page 109.*
*Oriental Asphalt Shingles, guaranteed 17 years, instead of wood shingles for roof, $29.00 extra.*
*Storm Doors and Windows, $99.00 extra.*
*Screen Doors and Windows, galvanized wire, $64.00 extra.*
*Oak Floors, Doors and Trim in Living Room and Dining Room on each floor and in Hall leading to second floor, also Stairs, Maple Floors in Kitchen and Bathroom, $312.00 extra.*

For prices of Plumbing, Heating, Wiring, Electric Fixtures and Shades see pages 130 and 131.

*For Our Easy Payment Plan See Page 144*

# Eight Rooms, Bath and Porches for Two Families

THE CLEVELAND is a two-story building with an apartment of four rooms and bathroom on each floor.

Each apartment is a complete home in itself, enjoying every privacy, thereby making a very attractive investment.

The general practice followed by the owner is to rent out one apartment and live in the other. Such rental generally pays the cost of the property and remains a source of income thereafter. Besides, there is nearly always a ready market for a two-apartment house of this kind. Profits of $1,000 and over are readily made.

**Exterior.** The Cleveland has the outward appearance of a one-family residence, yet each apartment has its own entrance. A porch on the first floor apartment and a balcony on the second floor apartment are features worth considering, not only because of their decorative value, but especially the delightful afternoons and evenings that can be leisurely enjoyed during the warm season each year.

**Can be built on a lot 28 feet wide.**

FIRST FLOOR PLAN

### Honor Bilt

## The Cleveland
No. P3233 "Already Cut" and Fitted
### $2,739.00

#### FIRST FLOOR APARTMENT

**The Living Room.** You enter from the spacious front porch into the living room. Here wall and floor space will accommodate a piano and living room furniture. A wide front window and a side window admit light and air. Floor dimensions, 11 feet 4 inches by 12 feet 5 in.

**The Kitchen.** A swinging door connects the living room and the kitchen. Size, 11 feet 4 inches by 11 feet 4 inches. The housewife will approve the arrangement of spaces for the range, the refrigerator, table and chairs; the built-in kitchen cabinet is located next to the sink, thus saving considerable time in the preparation of meals. A double window floods the kitchen with light and ventilation.

One door leads to hall connecting with bedrooms and bathroom. Another leads to side entry, which has stairs to basement and outdoors.

**The Bedrooms.** French doors open from the living room into the front bedroom. It has a clothes closet and is provided with cross ventilation. Floor space, 9 feet 6 inches by 10 feet 1 inch. A door leads to hall which connects with the rear bedroom, bathroom and kitchen.

The rear bedroom has a large clothes closet. Two windows assure cross ventilation. Size, 9 feet 6 inches by 10 feet.

**The Bathroom** measures 6 feet 1 inch by 5 feet 5 inches. All of the plumbing may be roughed in on one wall, saving on installation. It has a medicine case.

#### SECOND FLOOR APARTMENT

The rooms in this apartment are exactly the same as those of the first floor, except, however, entrance is made through the side entry into the kitchen.

**The Basement.** Room for furnace for both apartments or a furnace for each apartment; also, laundry and storage.

**Height of Ceilings:** First floor, 8 feet 6 inches from floor to ceiling. Second floor, 8 feet 6 inches from floor to ceiling. Basement, 7 feet from floor to joists.

### What Our Price Includes

At the price quoted we will furnish all of the material to build this two-story, two-family apartment house, consisting of:
Lumber; Lath;
Roofing, Best Grade Clear Red Cedar Shingles;
Siding, Clear Cypress or Clear Red Cedar Bevel;
Framing Lumber, No. 1 Quality Douglas Fir or Pacific Coast Hemlock;
Flooring, Clear Grade Douglas Fir or Pacific Coast Hemlock;
Porch Flooring, Clear Grade Douglas Fir or Pacific Coast Hemlock;
Porch Ceiling, Clear Grade Douglas Fir or Pacific Coast Hemlock;
Finishing Lumber;
High Grade Millwork (see pages 110 and 111);
Interior Doors, Two-Panel Design of Douglas Fir;
Trim, Beautiful Grain Douglas Fir or Yellow Pine;
Windows of California Clear White Pine;
Medicine Cases;
Building Paper; Sash Weights;
Eaves Trough and Down Spout;
Stratford Design Hardware (see page 132);
Paint for Three Coats for Outside Trim and Siding;
Stain for Two Brush Coats for Shingles on Gable Walls;
Shellac and Varnish for Interior Trim and Doors.
Complete Plans and Specifications;
Price does not include cement, brick or plaster.
See description of "Honor Bilt" Houses on pages 12 and 13.

SECOND FLOOR PLAN

#### OPTIONS

*Sheet Plaster and Plaster Finish to take the place of wood lath, $237.00 extra. See page 109.*

*Storm Doors and Windows, $89.00 extra.*

*Screen Doors and Windows, galvanized wire, $58.00 extra.*

*Oriental Slate Surfaced Shingles in place of wood shingles, $30.00 extra.*

For Prices for Plumbing, Heating, Wiring, Electric Fixtures and Shades see pages 130 and 131.

*For Our Easy Payment Plan See Page 144*

**Honor Bilt**

## The Betsy Ross

### No. P3089 "Already Cut" and Fitted

## $1,654.00

THE BETSY ROSS Home combines the charm of the pure colonial exterior with all of the comforts and economies of a bungalow. Judges of good architecture proclaim the perfect proportions and balance of this home.

**Exterior.** Briefly, here are just a few of the main points that particularly please owners of The Betsy Ross Home: (1) True colonial entrance; (2) wide siding, shutters and wood shingles; (3) red brick chimney with flower box. The brick terrace is a beauty mark in the center of which is the entrance hood and pillars. Flower trellises add the finishing touch to this perfectly balanced exterior.

**Interior.** The best of all good points of the interior is the arrangement of the rooms.

**The Living Room** is large and sunny. Size, 18 feet by 12 feet 5 inches. At the left end of this spacious 18-foot room is a handsome colonial mantel and fireplace. A coat closet is at the opposite end. Remarkably long wall spaces permit almost any arrangement of living room furniture. French windows on two sides furnish light and air.

**The Dining Room or Bedroom.** A hall connects the living room, dining room, kitchen and bathroom. The dining room or bedroom is 9 feet 5 inches by 10 feet 8 inches. It is generally used as a bedroom. A formal dining room in this home is considered unnecessary as there is a large breakfast alcove pleasantly situated right off the kitchen. It has a clothes closet; two windows provide light and cross ventilation.

**The Kitchen.** One likes to prepare the daily meals here. Work is done quickly. You will find the space for range is close to the sink, while our built-in cabinet, with its many labor saving conveniences, is directly opposite. Another feature that is greatly appreciated is the built-in ironing board. Both of the above mentioned built-in features are illustrated on pages 110 and 111. Light and cross ventilation. A door opens to lighted rear entry which has space for ice box, stairs to basement and door to grade.

**The Breakfast Alcove** connects with the kitchen through a wide cased opening. See illustration of breakfast alcove furniture on page 111.

**The Bedroom** at the front opens from the living room. It measures 11 feet 4 inches by 9 feet 9 inches. It has a clothes closet; two windows admit light and air.

**The Bathroom** has a built-in medicine case, and a handy closet for a broom, vacuum cleaner, etc.

**The Basement.** Excavated under entire house. Room for furnace, laundry and storage.

**Height of Ceilings.** Main floor, 9 feet from floor to ceiling. Basement, 7 feet from floor to ceiling.

### What Our Price Includes

At the price quoted we will furnish all the material to build this four-room bungalow consisting of:

**Lumber; Lath;**
**Roofing,** Best Grade Clear Red Cedar Shingles;
**Siding,** Clear Cypress or Clear Red Cedar, Bevel;
**Framing Lumber,** No. 1 Quality Douglas Fir or Pacific Coast Hemlock;
**Flooring,** Clear Douglas Fir or Pacific Coast Hemlock;
**Porch Ceiling,** Clear Douglas Fir or Pacific Coast Hemlock;

**Can be built on a lot 37 feet wide**

FLOOR PLAN

**Finishing Lumber;**
**High Grade Millwork** (see pages 110 and 111);
**Interior Doors,** Five Cross Panel Design of Douglas Fir;
**Trim,** Beautiful Grain Douglas Fir or Yellow Pine;
**French Windows,** California Clear White Pine;
**Medicine Case; Colonial Shutters;**
**Brick Mantel; Kitchen Cupboard;**
**Breakfast Alcove; Ironing Board;**
**Eaves Trough and Down Spout;**
**40-Lb. Building Paper; Sash Weights;**
**Stratford Design Hardware** (see page 132);
**Paint** for Three Coats Outside Trim and Siding;
**Shellac and Varnish** for Interior Trim and Doors.
Complete Plans and Specifications.
See description of "Honor Bilt" Houses on pages 12 and 13.
We guarantee enough material to build this house. Price does not include cement, brick or plaster.

### OPTIONS

*Sheet Plaster and Plaster Finish, to take the place of wood lath, $129.00 extra. See page 109.*

*Oriental Asphalt Shingles guaranteed 17 years, instead of wood shingles, $32.00 extra.*

*Storm Doors and Windows, $70.00 extra.*

*Screen Doors and Windows, galvanized wire, $41.00 extra.*

For prices of Plumbing, Heating, Wiring, Electric Fixtures and Shades see pages 130 and 131.

### For Our Easy Payment Plan See Page 144

## The Tarryton

### No. P3247 "Already Cut" and Fitted

## $2,967⁰⁰

THE TARRYTON two-story home is a modernized version of the ever popular English colonial cottage. Carefully designed as to interior arrangement and exterior decoration, it fulfills every expectation of a good home. A friendly front entrance, good window arrangement and shutters, colonial siding with wide exposure, all add to its permanent attractiveness. Makes a good investment and, if sold, usually nets a handsome profit.

**Can be built on a lot 31 feet wide**

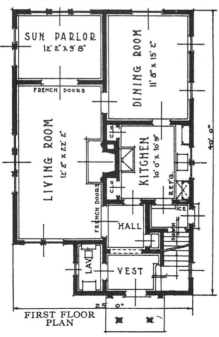

FIRST FLOOR PLAN

### FIRST FLOOR

**The Vestibule.** The entrance door opens into a vestibule which is separated from the hall by the use of a plaster arch with semi-circle head. On the right, a stairway leads to the second floor. On the left is the first floor lavatory, which is lighted by a window.

**The Living Room.** French doors connect the hall with the correctly proportioned living room. Size of living room, 12 feet 2 inches by 22 feet 2 inches. Here the inner wall is graced by a mantel and fireplace. Directly opposite the fireplace is a triple window through which the sun comes into the room to brighten the day. Liberal wall space and floor area will accommodate a piano and proper setting of furniture. A double window on the front adds additional light.

**The Sunroom.** Here, too, French doors connect the living room with the sunroom which, when open, give these rooms the appearance of one. Floor area, 12 feet 2 inches by 9 feet 8 inches. A triple window on the rear wall and a double window on the side wall floods the sunroom with sunshine.

**The Dining Room.** A single French door divides the living room and the dining room. Size of dining room is 11 feet 8 inches by 15 feet 2 inches. Lighted by windows on the side wall and a high sash on the rear wall.

**The Kitchen.** There is a swinging door between the dining room and the splendidly equipped kitchen. Every convenience needed to materially assist the housewife is provided in the layout of the space. For instance: The space for the range is located so that the flue can be placed in the fireplace chimney. On each side of the chimney is a closet which, when properly shelved, will act as a combination pantry and cupboard and also storage for brooms, vacuum, mop, etc. On the right wall the two windows are placed high up from the floor so that the kitchen sink can be placed under them. On one side of the sink a built-in pan and china cabinet are planned, while the other side is devoted to china cabinet with the space underneath for refrigerator, which can be iced from the side entry. A door leads to side entry which has stairs to basement and grade.

### SECOND FLOOR

**The Bedrooms.** The stairs from the vestibule lead to the long hall on the second floor. This hall connects with the three bedrooms, sewing room, bathroom, linen closet and stairway to attic. Each of the three bedrooms has three windows and a clothes closet. The sewing room is lighted by one window. The bathroom has a built-in medicine cabinet and a window.

**The Attic** can be used for various purposes.

**The Basement.** Room for heating plant, laundry and furnace.

**Height of Ceilings.** First floor, 9 feet from floor to ceiling. Second floor, 8 feet 6 inches from floor to ceiling. Basement, 7 feet from floor to joists.

### What Our Price Includes

At the price quoted we will furnish all the material to build this eight-room house, consisting of:

Lumber; Lath;
Roofing, Best Grade Clear Red Cedar Shingles;
Siding, Clear Cypress or Clear Red Cedar, Bevel;
Framing Lumber, No. 1 Quality Douglas Fir or Pacific Coast Hemlock;

Flooring, Clear Grade Douglas Fir or Pacific Coast Hemlock;
High Grade Millwork (see pages 110 and 111);
Interior Doors, Two-Panel Design of Douglas Fir;
Trim, Beautiful Grain Douglas Fir or Yellow Pine;
Windows of California Clear White Pine;
Medicine Case; Kitchen Cabinets;
Mantel; Colonial Shutters;
Eaves Trough and Down Spout;
40-Lb. Building Paper; Sash Weights;
Chicago Design Hardware (page 132);
Paint for Three Coats Outside Trim and Siding;
Varnish and Shellac for Interior Doors and Trim.

We guarantee enough material to build this house. Price does not include cement, brick or plaster.

See description of "Honor Bilt" Houses on pages 12 and 13 of Modern Homes Catalog.

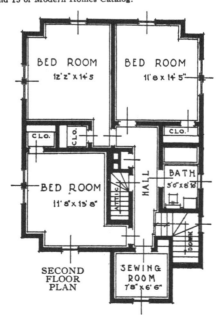

SECOND FLOOR PLAN

### OPTIONS

Sheet Plaster and Plaster Finish, to take the place of wood lath, $260.00 extra. See page 109.
Storm Doors and Windows, $139.00 extra.
Screen Doors and Windows, galvanized wire, $89.00 extra.
Oriental Slate Surfaced Shingles, in place of wood shingles, for roof, $43.00 extra.

For prices of Plumbing, Heating, Wiring, Electric Fixtures and Shades see pages 130 and 131.

*For Our Easy Payment Plan See Page 144*

THE WHITEHALL is a modern type of a two-story home that is seen in the better residential neighborhoods. It is dignified and substantial, and was designed with three important objects in view: Economy of floor space, quality of material and value-giving.

The Whitehall has been built in Gary, Ind., Allentown, Pa., Aurora, Ill., Plainville, Conn., McKeesport, Pa., Rich Valley, Ind., Hellertown, Pa., and other cities.

**Exterior.** The front elevation is pleasing because of the first and second floor bay windows, the dormer, the spacious porch and stairs and lattice work underneath. The porch measures 22 feet by 8 feet, and so, with ample space, family and guests can gather here, or children can play.

**Can be built on a lot 28 feet wide**

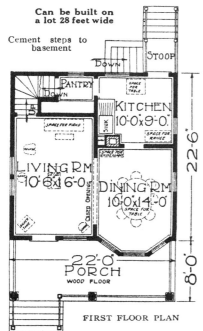

FIRST FLOOR PLAN

### Honor Bilt

## The Whitehall
### No. P3035A "Already Cut" and Fitted
## $1,863.00

#### FIRST FLOOR

**The Living Room,** 10 feet 6 inches by 16 feet, opens from the front porch. Directly opposite from the front entrance a door connects with the stairs that lead to the second floor.

The living room wall and floor space accommodates a piano and furniture. Light and air from windows on two sides.

**The Dining Room** and the living room make practically one large room because of the wide cased opening that connects them. The floor area measures 10 feet by 14 feet, thereby affording liberal space for the dining set. Three bay windows and one side window provide a flood of sunshine and air.

**The Kitchen** is entered from the dining room through a swinging door. It is at once the pride of the housewife. Everything she needs in the daily preparation of food is so handy! Here is space for the sink, range, table and refrigerator. A door connects with the pantry and stairs to basement. Another door leads to the rear stoop, which has stairs to grade, and outdoor entry to basement. The kitchen and the pantry are lighted and ventilated by windows.

#### SECOND FLOOR

**The Bedrooms.** The stairway from the living room leads to the second floor hall. This hall connects with the three bedrooms and the bathroom. Each bedroom has a clothes closet and is well lighted by windows.

**The Bathroom** plumbing can be roughed-in on one wall, thereby saving on installation. It has a built-in medicine case, and is lighted and aired by a window.

**The Basement.** Room for furnace, laundry and storage.

**Height of Ceilings.** First floor, 9 feet from floor to ceiling. Second floor, 8 feet 2 inches from floor to ceiling. Basement, 7 feet from floor to joists.

#### What Our Price Includes

At the price quoted we will furnish all the material to build this six-room house, consisting of:
**Lumber; Lath;**
**Roofing,** Best Grade Clear Red Cedar Shingles;
**Siding,** Clear Cypress or Clear Red Cedar, Bevel;
**Framing Lumber,** No. 1 Quality Douglas Fir or Pacific Coast Hemlock;
**Flooring,** Clear Douglas Fir or Pacific Coast Hemlock;
**Porch Flooring,** Clear Edge Grain Fir;
**Porch Ceiling,** Clear Douglas Fir or Pacific Coast Hemlock;

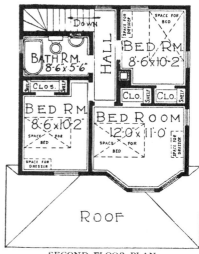

SECOND FLOOR PLAN

**Finishing Lumber;**
**High Grade Millwork** (see pages 110 and 111);
**Interior Doors,** Five-Cross Panel Design of Douglas Fir;
**Trim,** Beautiful Grain Douglas Fir or Yellow Pine;
**Windows,** California Clear White Pine;
**Medicine Case;**
**Eaves Trough and Down Spout;**
**Stratford Design Hardware** (see page 132);
**Paint** for Three Coats Outside Trim and Siding;
**Shellac and Varnish** for Interior Trim and Doors.

Complete Plans and Specifications.

We guarantee enough material to build this house. Price does not include cement, brick or plaster. See description of "Honor Bilt" Houses on pages 12 and 13.

#### OPTIONS

*Sheet Plaster and Plaster Finish, to take the place of wood lath, $177.00 extra. See page 109.*

*Oriental Asphalt Shingles, guaranteed 17 years, instead of wood shingles, $33.00 extra.*

*Oak Doors, Trim and Floors, for living room and dining room, Maple Floors in kitchen and bathroom, $112.00 extra.*

*Storm Doors and Windows, $77.00 extra.*

*Screen Doors and Windows, galvanized wire, $50.00 extra.*

For prices of Plumbing, Heating, Wiring, Electric Fixtures and Shades see pages 130 and 131.

*For Our Easy Payment Plan See Page 144*

**Honor Bilt**

*The Windermere*
No. P1208 Not Cut or Fitted
$3,534.00

THE WINDERMERE is a good two-family apartment house with five rooms and bathroom for each family. It makes a reliable and paying investment, increasing in value with the growth of the community.

**The Exterior.** Modern and attractive. The two large porches, size 26 feet by 8 feet, with their massive columns and solid buttress, makes glazing or screening in an inexpensive matter. The carefully planned roofs of house and porch, with their enclosed cornices and heavy verge boards, add to its character and stability. The upkeep will be small.

**FIRST AND SECOND FLOORS**
Both apartments are laid out identically the same with the exception of the entrance to the living room. On the first floor the entrance is directly from the large front porch, while on the second floor the entrance is from the stair hall.

**The Living Room.** Size. 13 feet 4 inches by 11 feet 4 inches. Space for piano and furniture. Has four windows.

**The Dining Room** connects with the living room by a wide cased opening. Size of the dining room, 14 feet 10 inches by 11 feet 2 inches. There is a recessed bay on the outer wall that has twin china cases which have leaded glass doors, and a hinged seat containing drawers in between.

**The Kitchen.** A swinging door connects the dining room with the bright kitchen. Our kitchen De Luxe Outfit, illustrated below, occupies the entire right wall. Light and ventilation from three windows. A door opens to rear entry which has stairs to grade door.

**The Bedrooms.** Two bedrooms, bathroom and linen closet connect with hallway that is open off the dining room. Each bedroom is of good size, has a clothes closet, and is lighted by two windows. The bathroom has a medicine case.

**The Basement.** Space for heating plants, laundry and storage.

**Height of Ceiling.** First floor and second floor height from floor to ceiling, 8 feet 6 inches. Basement height from floor to joists. 7 feet.

### What Our Price Includes

At the price quoted we will furnish all the material to build this 10-room two-family apartment house, consisting of:

**Lumber; Lath;**
**Roofing,** Best Grade Clear Red Cedar Shingles;
**Siding,** Clear Cypress or Clear Red Cedar, Bevel, for First Floor, Clear Red Cedar Shingles for Second Floor;
**Framing Lumber,** No. 1 Quality Douglas Fir or Pacific Coast Hemlock;
**Flooring,** Clear Douglas Fir or Pacific Coast Hemlock;
**Porch Flooring,** Clear Edge Grain Fir;
**Porch Ceiling,** Clear Douglas Fir or Pacific Coast Hemlock;
**Finishing Lumber;**
**High Grade Millwork** (see pages 110 and 111);
**Interior Doors,** Two-Panel Design Douglas Fir;

**Trim,** Beautiful Grain Douglas Fir or Yellow Pine;
**Windows** of California Clear White Pine;
**Medicine Cases;**
**China Cabinets;**
**Built-In Seats;**
**De Luxe Tile Sink Outfits;**
**Eaves Trough and Down Spout;**
**40-Lb. Building Paper; Sash Weights;**
**Stratford Design Hardware** (see page 132);
**Paint** for Three Coats Outside Trim and Siding;
**Stain** for Shingles on Walls for Two Brush Coats;
**Shellac and Varnish** for Interior Trim and Doors.

Complete Plans and Specifications.
We guarantee enough material to build this house. Price does not include cement, brick or plaster.
See description of "Honor Bilt" Houses on pages 12 and 13.

### OPTIONS

*Sheet Plaster and White Plaster Finish, in place of wood lath,* $313.00 *extra. See page 109.*

*Oak Floors, Doors, Trim for living room and dining room, Maple Floors in kitchen and bathroom,* $296.00 *extra.*
For prices of Plumbing, Heating, Wiring, Electric Fixtures and Shades see pages 130 and 131.
*Two furnaces are included when we furnish this building with warm air heat, but when furnished with*

*Oriental Asphalt Shingles, guaranteed 17 years, for roof, instead of wood shingles,* $62.00 *extra.*
*Storm Doors and Windows,* $130.00 *extra.*
*Screen Doors and Windows, galvanized wire,* $88.00 *extra.*

*hot water or steam heating only one boiler is furnished to heat both apartments.*

### For Our Easy Payment Plan See Page 144

**Can be built on a lot 35 feet wide**

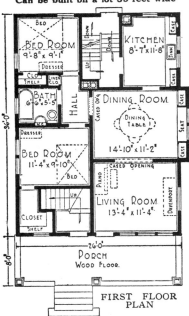

FIRST FLOOR PLAN

SECOND FLOOR PLAN

THE BEDFORD bungalow home is readily admired wherever built. Its exterior is given a substantial appearance by the use of heavy columns and solid porch railing. It is planned to be constructed with first story of brick veneer and with shingle gables and dormer.

### THE FIRST FLOOR

**The Living Room.** The big front porch, size 25 feet by 8 feet, leads to the entrance hall.

The living room measures 19 feet 1 inch by 12 feet 5 inches and so makes an ideal room. The left wall is dignified by a mantel and fireplace, which has a high casement window on each side. Generous wall space and floor area provide for accommodating a piano and complete furnishings. The attractive grouping of four front windows gives the room plenty of light.

**Can be built on a lot 31 feet wide**

FIRST FLOOR PLAN

**The Dining Room.** The opening between the living room and the dining room is equipped with French doors. The dining room has a bay window extension, with triple windows, thereby adding to the attractiveness of social occasions. Floor space, 13 feet 3 inches by 13 feet 7 inches.

**The Kitchen** is entered from the dining room through a swinging door. Much thought has been given to provide a convenient place for all necessary kitchen equipment. For example: The sink is located under the double windows, with built-in cabinets and pan cupboards on each side. The range and refrigerator space is located on the right wall. A door leads to the rear entry, which has stairs to grade and basement.

**The Bedrooms** on this floor connect with the small hall immediately off the dining room. Each bedroom has two windows and a clothes closet.

**The Bathroom** is located between the bedrooms. All of the plumbing may be roughed-in on one wall saving on installation cost. It has a built-in medicine cabinet and a window.

### THE SECOND FLOOR

The stairway from the first floor leads to the hall on the second floor. This hall connects with the three bedrooms. The front bedroom is generally used as an upstairs sewing room or living room; size 11 feet 2 inches by 12 feet 5 inches. The bedroom at the left is 11 feet 11 inches by 16 feet 4 inches, and the bedroom at the right measures 11 feet 11 inches by 12 feet 6 inches. Each of the rear bedrooms has a good size clothes closet.

The arrangement of the upstairs bedrooms, separated as they are from the rooms on the first floor, is very practical from a revenue earning standpoint. The upstairs bedrooms can be used from the outside without going through the living rooms of the first floor. Then, again, it makes a good home for large families.

**The Basement.** Room for heating plant, laundry and furnace.

**Height of Ceilings.** First floor, 9 feet from floor to ceiling. Second floor, 8 feet 6 inches from floor to ceiling. Basement, 7 feet from cement floor to joists.

### What Our Price Includes

At the price quoted we will furnish all the material to build this eight-room bungalow, consisting of:
Lumber; Lath;
Roofing, Best Grade Clear Red Cedar Shingles;
Siding, Clear Cypress or Clear Red Cedar, Bevel;
Framing Lumber, No. 1 Quality Douglas Fir or Pacific Coast Hemlock;
Flooring, Clear Grade Douglas Fir or Pacific Coast Hemlock;
Porch Ceiling, Clear Grade Douglas Fir or Pacific Coast Hemlock;
Finishing Lumber;
High Grade Millwork (see pages 110 and 111);
Interior Doors, Two-Panel Design Douglas Fir;
Trim, Beautiful Grain Douglas Fir or Yellow Pine;
Windows of California Clear White Pine;
Medicine Case; Mantel;
Kitchen De Luxe Outfit;
Eaves Trough and Down Spout;
40-Lb. Building Paper; Sash Weights;
Chicago Design Hardware (see page 132);
Paint for Three Coats Outside Trim;
Stain for Shingles on Gables for Two Brush Coats;
Varnish and Shellac for Interior Doors and Trim

We guarantee enough material to build this house. Price does not include cement, brick or plaster.

See description of "Honor Bilt" Houses on pages 12 and 13 of Modern Homes Catalog.

## The *Bedford*
### No. P3249 "Already Cut" and Fitted
### $2,396.00

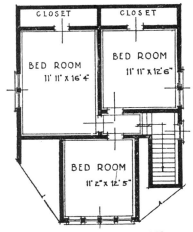

SECOND FLOOR PLAN

### OPTIONS

*Sheet Plaster and Plaster Finish, to take the place of wood lath, $260.00 extra. See page 109.*

*Storm Doors and Windows, $118.00 extra.*

*Screen Doors and Windows, galvanized wire, $77.00 extra.*

*Oriental Slate Surfaced Shingles, in place of wood shingles, $49.00 extra.*

For prices of Plumbing, Heating, Wiring, Electric Fixtures and Shades see pages 130 and 131.

*For Our Easy Payment Plan See Page 144*

## The Manchester
### No. P3250 "Already Cut" and Fitted
## $2,655.00

THE MANCHESTER Home has the appearance of a one-family, two-story modern bungalow. Careful planning down to the smallest detail makes The Manchester a very good example of the so called "Income" bungalow. The first floor accommodates a 5-room apartment with bathroom, and the second floor accommodates a 4-room apartment with bathroom. This type of building not only provides a fine home for the owner but also furnishes a comfortable income by renting the other apartment to a tenant, whose rental usually pays for the cost of the investment. Again, the cost of constructing a home of this type is very little more than the cost of a one-story or a story and a half bungalow.

### FIRST FLOOR APARTMENT

**The Porch** measures 24 feet by 8 feet.

**The Entrance** opens from the front porch into a small hall, which contains stairs leading to the second floor apartment.

**The Living Room** is 19 feet 2 inches by 11 feet 5 inches. The mantel and fireplace, with high casement windows on each side, is on the left wall. Space underneath the high casement windows can be used for bookcases or other convenient furniture. Additional wall space will accommodate a piano, while floor area permits furnishing to please. Four front windows flood the living room with sunshine.

**The Dining Room.** The opening between the living room and the dining room is planned for a coved arch. The dining room is 12 feet 5 inches by 11 feet 7 inches and is well lighted by a double window at the side.

**The Kitchen** is entered from the dining room through a swinging door. Though small, size 8 feet 3 inches by 9 feet 9 inches, it provides a convenient place for sink, range, refrigerator and built-in cupboard. A double window admits light and ventilation. A door connects with the rear entry, which has stairs to basement and grade.

**The Bedrooms.** A small hall opening off the dining room provides privacy to the bedrooms and bathroom. The center bedroom measures 10 feet 5 inches by 11 feet 7 inches. Contains a double window, and a clothes closet, located under the main stairs. The rear bedroom has a large clothes closet and two windows with cross ventilation.

**The Bathroom** plumbing can be roughed in on one wall, saving on installation cost; has a built-in medicine cabinet and a window.

The location of the kitchen and bathroom on the first floor under practically the same space of the kitchen and bathroom on the second floor reduces the plumbing installation down to the lowest cost.

### SECOND FLOOR APARTMENT

**The Living Room.** The stairway from the first floor hall leads to the small hall on the second floor that connects with the good size living room: 15 feet 2 inches by 13 feet 9 inches. It has a large closet that will accommodate a folding bed, which gives this apartment a two-bedroom efficiency. Three front windows admit light and air.

**The Dining Room** measures 11 feet 5 inches by 10 feet 1 inch. It has a closet, and is lighted by a window.

**The Kitchen.** A swinging door connects the dining room with the kitchen. Size of kitchen, 9 feet 8 inches by 9 feet 5 inches. Space is provided for the range, refrigerator and sink. Two windows over the place for sink provide light and ventilation. Close to the sink is a built-in kitchen cupboard. A door leads to the rear porch, which has an open stairway to grade, and basement.

**The Bedroom** is entered from the living room or from the hall off the kitchen which connects, also, with the bathroom. The bedroom has a clothes closet, and a window.

**The Bathroom** has a built-in medicine cabinet, and a window.

**The Basement.** Room for furnace, laundry and storage.

**Height of Ceilings.** First floor, 9 feet from floor to ceiling. Second floor, 8 feet 6 inches from floor to ceiling. Basement, 7 feet from floor to joists.

### OPTIONS

*Sheet Plaster and Plaster Finish, to take the place of wood lath, $261.00 extra. (See page 109).*
*Storm Doors and Windows, $126.00 extra.*
*Screen Doors and Windows, galvanized wire, $87.00 extra.*

### For Our Easy Payment Plan See Page 144

For prices of Plumbing, Heating, Wiring, Electric Fixtures and Shades see pages 130 and 131.

### What Our Price Includes

At the price quoted we will furnish al the material to build this two-family, nine-room bungalow, consisting of:
Lumber; Lath;
Roofing, Oriental Slate Surfaced Shingles;
Siding, Clear Cypress or Clear Red Cedar Bevel;
Framing Lumber, No. 1 Quality Douglas Fir or Pacific Coast Hemlock;
Flooring, Clear Grade Douglas Fir or Pacific Coast Hemlock;
Porch Flooring, Clear Grade Douglas Fir or Pacific Coast Hemlock;
Porch Ceiling, Clear Grade Douglas Fir or Pacific Coast Hemlock;
Finishing Lumber; High Grade Millwork (see pages 110 and 111);
Interior Doors, Two-Panel Design Douglas Fir;
Trim, Beautiful Grain Douglas Fir or Yellow Pine;
Windows of California Clear White Pine;
Medicine Case; Kitchen Cabinet; Mantel;
Eaves Trough and Down Spout;
40-Lb. Building Paper; Sash Weights;
Chicago Design, Hardware (see page 132);
Paint for Three Coats Outside Trim and Siding;
Stain for Shingles on Gables for Two Brush Coats;
Varnish and Shellac for interior Doors and Trim.

We guarantee enough material to build this house. Price does not include cement, brick or plaster.

See description of "Honor Bilt" Houses on pages 12 and 13 of Modern Homes Catalog.

**Can be built on a lot 30 feet wide**

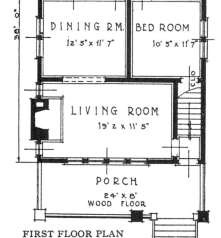

FIRST FLOOR PLAN

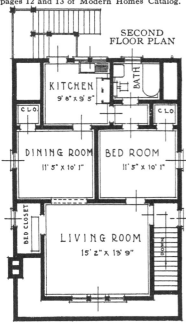

SECOND FLOOR PLAN

## The Winona

**$1,742.00** No. P2010 "Already Cut" and Fitted. Five Rooms.
**$1,985.00** No. P2010B "Already Cut" and Fitted. Six Rooms.

THE WINONA bungalow comes either in five or six rooms, bath and large front porch. Either size is a big value for the money. Built in many places. Satisfied owners are full of praise.

Every bit of material is "Honor Bilt," and comes direct from our large manufacturing plants.

**The Exterior.** Note the gracefully paneled effect on the porch columns and along the sides of the bungalow. Our Oriental Sea Green Slate Surfaced Siding is used in the panels shown in the illustration above. The balance of the building is sided with best grade thick red cedar shingles.

**The Porch** measures 24 feet by 8 feet. The family, especially the children, will enjoy the porch throughout the good weather months.

**The Living Room.** A graceful front door, glazed with large bevel plate glass, leads from the porch into the vestibule, which is connected with the living room by a cased opening. In design No. P2010B the front door leads directly into the living room. The size of each living room is given in the floor plan.

**The Dining Room** has a handsome sideboard in No. P2010, while in the other design there is a recessed double bay window and a hinged seat instead.

**The Kitchen** has a pantry in design No. P2010, and a built-in cabinet in No. P2010B. Of course, there is the usual space for the sink, range, table and chair—all conveniently arranged. Well lighted and aired. Door to grade, and stairs to basement.

**The Bedrooms.** In design No. P2010B there are three bedrooms, and in No. P2010 there are only two bedrooms. The bedrooms and the bathroom are entered from an open hall off the dining room. Each bedroom has two windows, and a clothes closet.

**The Basement.** Room for furnace, laundry and storage.

**Height of Ceilings.** Main floor, 9 feet from floor to ceiling. Basement, 7 feet from floor to joists.

### What Our Prices Include

At the prices quoted we will furnish all the material to build this five or six-room bungalow, consisting of:

Lumber; Lath;
Roofing, Best Grade Clear Red Cedar Shingles;
Siding, Best Grade Clear Red Cedar Shingles; Oriental Slate Surfaced Siding;
Framing Lumber, No. 1 Quality Douglas Fir or Pacific Coast Hemlock;
Flooring, Clear Douglas Fir or Pacific Coast Hemlock;
Porch Flooring, Clear Edge Grain Fir;
Porch Ceiling, Clear Douglas Fir or Pacific Coast Hemlock;
Finishing Lumber;
High Grade Millwork (see pages 110 and 111);
Interior Doors, Two Cross Panel Design for No. P2010B and Five Cross Panel Design for No. P2010, All of Douglas Fir;
Trim, Beautiful Grain Douglas Fir or Yellow Pine;
Windows, California Clear White Pine;
Medicine Case;
Kitchen Cabinet for No. P2010B;
Sideboard for No. P2010;
Eaves Trough and Down Spout;
40-Lb. Building Paper; Sash Weights;
Stratford Design Hardware (see page 132);
Paint for Three Coats Outside Trim;
Stain for Two Brush Coats for Shingles on Wall;
Shellac and Varnish for Interior Trim and Doors.

Complete Plans and Specifications.

We guarantee enough material to build this bungalow. Price does not include cement, brick or plaster. See description of "Honor Bilt" Houses on pages 12 and 13.

This bungalow can be built with the rooms reversed. See page 3.

**Can be built on a lot 32 feet wide**

FLOOR PLAN No. P2010B

No. P2010B has stairs to the attic, which is floored, 12 feet wide and extends the entire length of the house.

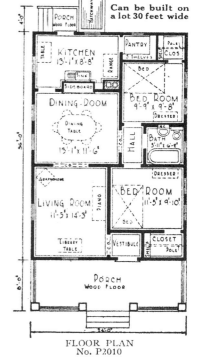

**Can be built on a lot 30 feet wide**

FLOOR PLAN No. P2010

### OPTIONS FOR No. P2010 AND No. P2010B

*Sheet Plaster and Plaster Finish, to take the place of wood lath, for No. P2010, $145.00 extra; for No. P2010B, $166.00 extra. See page 109.*

*Oriental Asphalt Shingles, guaranteed 17 years, instead of wood shingles, for roof No. P2010, $42.00 extra; for No. P2010B, $52.00 extra.*

*Oak Doors, Trim and Floors in living room and dining room, Maple Floors in kitchen and bathroom,*

*for No. P2010, $135.00 extra; for No. P2010B, $130.00 extra.*

*Storm Doors and Windows, No. P2010, $52.00 extra; for No. P2010B, $62.00 extra.*

*Screen Doors and Windows, No. P2010, galvanized wire, $33.00 extra; for No. P2010B, galvanized wire, $39.00 extra.*

For prices of Plumbing, Heating, Wiring, Electric Fixtures and Shades see pages 130 and 131.

### For Our Easy Payment Plan See Page 144

THE ROCKFORD two-story home is finished with veneer of brick. Substantial strength is expressed in its dignified exterior. The roof is of a hipped type and any suggestion of plainness is eliminated by the use of a dormer in the front elevation, the tall brick fireplace chimney and the porch trellis for climbing plants. The solid brick rail gives the porch added privacy, thus increasing its usefulness to the family.

The Rockford is conveniently planned to allow the greatest use of space consistent with good architecture. All of the rooms are properly proportioned to give greater comfort and lasting satisfaction. Low construction cost is another favorable feature.

### FIRST FLOOR

**The Porch** measures 25 feet by 8 feet.

**The Hall.** The entrance is into a hall. This hall contains semi-open stairs that lead to second floor, a guest's clothes closet on the stair landing, a door to side entry that connects with the kitchen, and on the left is a well balanced arched opening to the living room.

**Can be built on a lot 31 feet wide**

**FIRST FLOOR PLAN**

**The Living Room.** Floor dimensions are 15 feet 5 inches by 13 feet 5 inches. Directly opposite the arched opening from the hall is a mantel and fireplace, with a high sash on each side. A triple window is in the front. Wall space and floor space accommodate a piano and furniture.

**The Dining Room.** Here, also, is an arched opening that serves to divide the living room and the dining room. Size of dining room, 12 feet 5 inches by 13 feet 5 inches. The triple high sash casement windows in the side wall are spaced so as to permit the placing of a buffet directly underneath. A double window is in the rear wall.

**The Kitchen** and the dining room are connected by a swinging door. Handily and efficiently arranged and so will delight the housewife. Immediately inside the door is space for range, and directly opposite underneath a double window is space for a sink. Then, again, on each wall opposite the sink is a built-in cupboard. At the very rear is a door to an open porch. On the right is a door that connects with the side entry which has space for a refrigerator. The side entry has stairs to basement and grade.

### SECOND FLOOR

**The Bedrooms.** A short hall connects with the three bedrooms, bathroom, and stairs leading to the first floor. Each bedroom has a clothes closet, the two bedrooms towards the left have three windows providing cross ventilation, while bedroom to the right has two windows.

**The Bathroom** has a built-in medicine case and a window.

**The Basement.** Room for furnace, laundry and storage.

**Height of Ceilings.** First floor, 9 feet from floor to ceiling. Second floor, 8 feet 8 inches from floor to ceiling. Basement, 7 feet from floor to joists.

### What Our Price Includes

At the price quoted we will furnish all the material to build this six-room house, consisting of:
Lumber; Lath;
Roofing, Best Grade Clear Red Cedar Shingles;
Framing Lumber, No. 1 Quality Douglas Fir or Pacific Coast Hemlock;
Flooring, Clear Grade Douglas Fir or Pacific Coast Hemlock;
Porch Ceiling, Clear Grade Douglas Fir or Pacific Coast Hemlock;
Finishing Lumber;
High Grade Millwork (see pages 110 and 111);
Interior Doors, Two-Panel Design Douglas Fir;
Trim, Beautiful Grain Douglas Fir or Yellow Pine;
Windows of California Clear White Pine;
Medicine Case; Kitchen Cabinets;
Mantel;
Eaves Trough and Down Spout;
40-Lb. Building Paper; Sash Weights;
Chicago Design Hardware (see page 132);
Paint for Three Coats Outside Trim;
Varnish and Shellac for Interior Doors and Trim.

We guarantee enough material to build this house. Price does not include cement, brick or plaster. See description of "Honor Bilt" Houses on pages 12 and 13 of Modern Homes Catalog.

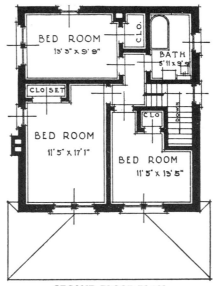

**Honor Bilt**

## The Rockford

### No. P3251 "Already Cut" and Fitted
### $2,278.00

**SECOND FLOOR PLAN**

### OPTIONS

*Sheet Plaster and Plaster Finish, to take the place of wood lath, $211.00 extra. See page 109.*

*Storm Doors and Windows, $118.00 extra.*

*Screen Doors and Windows, galvanised wire, $80.00 extra.*

*Oriental Slate Surfaced Shingles, in place of wood shingles, $43.00 extra.*

For prices of Plumbing, Heating, Wiring, Electric Fixtures and Shades see pages 130 and 131.

### For Our Easy Payment Plan See Page 144

THE ARDARA bungalow possesses many attractive features and conveniences. Here the architect has created a design that is conspicuous because of its beauty and notable for its contrast to the commonplace. It pleasingly combines oriental and occidental architecture. The porch, with its graceful roof lines, together with the trellises over the large front windows is in perfect harmony with rest of the design. The garage at the left, carrying the same treatment as the house, not only is a very convenient feature, but enhances the beauty of the entire bungalow. The Ardara has been chosen by many professional and business men because of its practical arrangement.

**Honor Bilt**

## The Ardara

### No. P13039 "Already Cut" and Fitted
**$2,245.00** **$2,483.00**
Without Garage With Garage

**The Living Room.** One enters the living room through a beautiful glazed door with side lights. The first impression is the large size of the rooms. The living room, dining room, and music room are separated by wide cased openings, giving one a feeling of spacious living quarters. The living room, with music room, is 24 feet 10 inches long by nearly 14 feet wide. This arrangement allows space for piano, davenport and all other living room furniture. A coat closet is conveniently located in the living room, close to the passage that leads to dining room and garage. There are four windows, giving an abundance of light and air.

The music room may also be used for library, study or professional office. It is located with the least interference to the rest of the house.

**The Dining Room.** The feature of the dining room is the triple window. The size of the room is 12 feet 6 inches by 13 feet 6 inches, allowing plenty of room for furniture and service.

**The Kitchen.** From dining room a swinging door leads to the well arranged kitchen, 10 feet 2 inches wide and 10 feet 7 inches deep. Space for range is just inside door. Sink is located in the left hand corner. Work table can be placed directly under the double windows at the right. Directly off the kitchen is a pantry, provided with shelves. A door leads to a rear entry. There is space for ice box. Rear door opens to stoop, which leads to lawn.

**The Bedrooms.** The arrangement of sleeping quarters and bath deserves special mention. The two bedrooms and bath are reached through a hallway that also connects with the dining room and kitchen. One bedroom is 12 feet 4 inches by 11 feet 4 inches. The rear bedroom is 11 feet 2 inches by 9 feet 10 inches. Bedrooms have good size clothes closets, with shelves and wardrobe poles. Both bedrooms have two windows. The bathroom is conveniently located to bedrooms and kitchen. Ample space is provided for bathroom fixtures. Due to the arrangement of the kitchen and bathroom, considerable saving is made in the "roughing in" of the plumbing.

**The Garage.** The attachment of the garage to the house is ideal. Many of the most expensive mansions are now arranged in this manner. Aside from the convenience and safety, the garage can be heated at a lower cost. A garage with house can be built more economically than if a

substantial garage were built separately. The garage connects with the living quarters through a hall, avoiding the usual discomforts when the garage is separate. The garage has large swinging doors in the front, two windows at the side and one window and grade entrance at the rear. Size of garage is 12 feet by 18 feet. The Ardara can be furnished with or without garage.

**Basement.** Entrance to basement is through the front hall. Room for furnace, laundry and storage.

**Height of Ceilings.** Basement, 7 feet, floor to joists. Main floor, 9 feet 2 inches, floor to ceiling.

### What Our Price Includes

At the price quoted we will furnish all the material to build this six-room house and garage, consisting of:
Lumber; Lath;
**Roofing,** Oriental Slate Surfaced Shingles, guaranteed 17 years;
**Siding,** Clear Cypress or Clear Red Cedar, Bevel;
**Framing Lumber,** No. 1 Quality Douglas Fir or Pacific Coast Hemlock;
**Flooring,** Clear Maple for Kitchen, Hall, Bathroom, Entry and Pantry; Clear Oak for Balance of Rooms;
**Porch Ceiling,** Clear Douglas Fir or Pacific Coast Hemlock;
**Finishing Lumber;**
**High Grade Millwork** (see pages 110 and 111);
**Interior Doors,** Five Cross Panel Design Douglas Fir;
**Trim,** Beautiful Grain Douglas Fir or Yellow Pine;
**Medicine Case;**
**Windows,** California Clear White Pine;
**40-Lb. Building Paper; Sash Weights;**
**Eaves Trough and Down Spout;**
**Stratford Design Hardware** (see page 132);
**Paint for Three Coats** Outside Trim and Siding;
**Shellac and Varnish** for Interior Trim and Doors;
**Shellac, Paste Filler and Floor Varnish** for Oak and Maple Floors;
Complete Plans and Specifications.
Built on a concrete foundation and excavated under entire house.
We guarantee enough material to build this house. Price does not include cement, brick or plaster.
See description of "Honor Bilt" Houses on pages 12 and 13.

#### OPTIONS

*Sheet Plaster and Plaster Finish, to take the place of wood lath, $199.00 extra. See page 109.*
*Storm Doors and Windows, $76.00 extra.*
*Screen Doors and Windows, galvanized wire, $49.00 extra.*
*Oak Doors and Trim for living room, music room and dining room, $135.00 extra.*
For prices on Plumbing, Heating, Wiring, Electric Fixtures and Shades see pages 130 and 131.

**Can be built on a lot 50 feet wide**

**FLOOR PLAN**

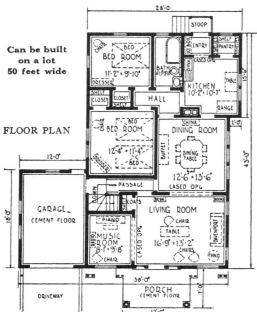

*For Our Easy Payment Plan See Page 144*

THE ARGYLE is a bungalow home that will not be too extreme and yet is entirely different from a cottage. The exterior is finished in shingles, except the gables and porch which call for stucco. It is neat, well arranged and solidly constructed. We have included the most popular built-in fixtures, thus saving both room and the need of purchasing bulky pieces of furniture, such as bookcases and kitchen cabinets. Moreover, careful study of the Argyle floor plan reveals as much actual accommodation and more convenience than the usual six or seven-room two-story house.

Argyle owners are very enthusiastic. Their letters freely praise our free architectural service, good material, solid construction and money saved on their houses. "A $7,500.00 house," you might say, and you would be right if it were built in the ordinary way. Yet, by our "Honor Bilt" System, we are able to furnish the materials so you can build The Argyle for a great deal less.

**The Living Room.** From the front porch, with its bungalow porch rail, you enter the living room. It is 12 feet 2 inches wide and 15 feet 11 inches long. A fine craftsman brick mantel sits in the center of the right wall. On each side of mantel is a built-in bookcase, glazed with leaded glass doors. A casement sash, corresponding in style with top of big front window, is directly above each bookcase. There is ample wall space for furniture and piano. Light and ventilation from two sides.

**The Dining Room.** You pass through a wide cased opening from the living room into the dining room, size 14 feet 2 inches by 11 feet 4 inches. Here the walls are paneled. Four windows in a recessed bay insure a cheerful atmosphere that adds zest when the family dines.

**The Kitchen.** A swinging door leads from the dining room to the ideal kitchen. It is 12 feet 2 inches by 9 feet 2 inches in size. On each side of space for sink are upper and lower cupboards. A complete cabinet is built on the opposite wall. There is ample space for a stove, table and other needed furniture. Three windows provide light and air. The grade entrance keeps cold and dirt out. Stairs lead to yard and basement.

**The Bedrooms.** A hall opens from the dining room and connects with the two bedrooms and bath. Hall has a roomy coat closet and also a linen closet. The front bedroom has a clothes closet with shelf. There is a front and also a side window. The rear bedroom, too, has a clothes closet with hat shelf. There are two windows on the side. Bathroom is conveniently located between bedrooms.

## The Argyle

### No. P17018A "Already Cut" and Fitted
### $2,150.00

**Basement.** Excavated basement with concrete floor. Room for furnace, laundry and storage.

**Height of Ceilings.** Main floor, 8 feet 2 inches from floor to ceiling. Basement, 7 feet high from floor to joists.

## What Our Price Includes

At the price quoted we will furnish all the material to build this five-room house, consisting of:
Lumber; Lath;
**Roof Shingles,** Best Grade Thick Cedar;
**Siding,** Best Grade Thick Cedar Shingles;
**Framing Lumber,** No. 1 Quality Douglas Fir or Pacific Coast Hemlock;
**Flooring,** Clear Maple for Kitchen and Bathroom; Clear Oak for Other Rooms; Fir for Porch;
**Porch Ceiling,** Clear Douglas Fir or Pacific Coast Hemlock;
**Finishing Lumber;**
**High Grade Millwork** (see pages 110 and 111);
**Interior Doors,** Two Vertical Panel Design of Douglas Fir;
**Trim,** Beautiful Grain Douglas Fir or Yellow Pine;
**Kitchen Cupboards;**
**Medicine Case;**
**Brick Mantel;**
**Windows,** California Clear White Pine;
**40-Lb. Building Paper; Sash Weights;**
**Eaves Trough and Down Spout;**
**Chicago Design Hardware** (see page 132);
**Paint** for Three Coats Outside Trim;
**Stain** for Shingles on Walls for Two Brush Coats;
**Shellac and Varnish** for Interior Trim and Doors;
**Shellac, Paste Filler and Floor Varnish** for Oak and Maple Floors.

Complete Plans and Specifications.
Built on concrete foundation and excavated under entire house.
We guarantee enough material to build this house. Price does not include cement, brick or plaster.

See description of "Honor Bilt" Houses on pages 12 and 13.

**Can Be Built on 33-Foot Lot**

This house can be built with the rooms reversed. See page 3.

FLOOR PLAN

### OPTIONS

*Sheet Plaster and Plaster Finish to take the place of wood lath, $153.00 extra. See page 109.*

*Oriental Asphalt Shingles, guaranteed 17 years, instead of wood shingles for roof, $44.00 extra.*

*Oak Doors and Trim in living room and dining room, $76.00 extra.*

*Storm Doors and Windows, $64.00 extra.*

*Screen Doors and Windows, galvanized wire, $38.00 extra.*

For prices of Plumbing, Heating, Wiring, Electric Fixtures and Shades, see pages 130 and 131.

*For Our Easy Payment Plan See Page 144*

**Honor Bilt**

## The Sunlight
### No. P3221 "Already Cut" and Fitted
### $1,620.00

IN THIS modern five-room bungalow the architects have carefully planned every detail, that every inch of space is used to the best advantage. A careful study of the floor plan will reveal that the arrangement is ideal in every particular, resulting in the greatest amount of comfort, the lowest cost of fuel and minimum cost of upkeep. The careful planning of the "Sunlight" relieves the usual household drudgery. The high quality materials are the same as in all "HONOR BILT" homes. The low price is due: First, to the careful thought in its planning, and second, to the fact that the materials are figured at factory prices.

Front and rear gables ornamented with wood shingles, which can be stained in a pleasing tone. Porch, 24 by 8 feet, protects the front windows and door from snow, rain and sun. It makes an ideal place to enjoy the pleasant weather. Here is room for porch swing and furniture. A nice place for the kiddies to play. An enclosed rear entry is a feature.

**The Living Room.** Three steps lead to the front porch, which opens into the living room through an eight-light panel door. The arrangement of the floor plan permits either a combination living room and dining room, or separate rooms. The living room, size 12 feet 8 inches wide by 12 feet 2 inches long, has space for piano, davenport and other furniture. Windows on two exposures provide plenty of light and ventilation.

**The Dining Room.** A wide cased opening leads from the living room into dining room. Here a buffet may be placed on the inside wall. Here the family may dine in a cheerful atmosphere. Double windows supply an abundance of light and fresh air.

**The Kitchen.** A swinging door leads from the dining room into the kitchen. Here the architect has considered the daily tasks of the housewife. The range space and sink are so arranged as to take all of the "backaches" out of the work. Near by is a convenient cupboard for china, glassware and utensils. Plenty of air and light is provided by two windows. The rear door leads to an enclosed entry, with stairway to basement, and outside entrance. Space is provided for refrigerator.

**The Bedrooms.** Passing from the dining room, you enter a hall that connects with the two bedrooms and bathroom. Directly off the hall is a linen closet. The front bedroom is of good size. A clothes closet is provided with a shelf and wardrobe pole. There is a rear bedroom, size 10 feet 2 inches by 10 feet, with clothes closet. Each bedroom has two windows, giving ample light and ventilation. The arrangement of bathroom provides for tub in a recess, toilet and lavatory.

**The Basement.** Space for laundry, storage rooms and fuel bins.

**Height of Ceilings.** Basement, 7 feet high from floor to ceiling. Main floor, 9 feet from the floor to ceiling.

### What Our Price Includes

**At the price quoted we will furnish all the material to build this five-room bungalow, consisting of:**

Lumber; Lath;
**Roofing,** Best Grade Clear Red Cedar Shingles;
**Siding,** Clear Cypress, or Clear Red Cedar, Bevel;
**Framing Lumber,** No. 1 Quality Douglas Fir or Pacific Coast Hemlock;
**Flooring,** Clear Douglas Fir or Pacific Coast Hemlock;
**Porch Flooring,** Clear Edge Grain Fir;
**Porch Ceiling,** Clear Grade Douglas Fir or Pacific Coast Hemlock;
**Finishing Lumber;**
**High Grade Millwork** (see pages 110 and 111);
**Interior Doors,** Two Cross Panel Design of Douglas Fir;
**Trim,** Beautiful Grain Douglas Fir or Yellow Pine;
**Medicine Case;**
**Windows of California White Pine;**
**40-Lb. Building Paper; Sash Weights;**
**Eaves Trough; Down Spout;**
**Stratford Design Hardware** (see page 132);
**Paint** for Three Coats Outside Trim and Siding;
**Stain** for Shingles of Gables for Two Brush Coats;
**Shellac and Varnish** for Interior Trim and Doors.

Complete Plans and Specifications.

Built on concrete foundation and excavated under entire house.

We guarantee enough material to build this house. Price does not include cement, brick or plaster. See description of "Honor Bilt" Houses on pages 12 and 13.

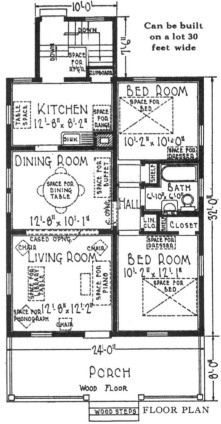

**Can be built on a lot 30 feet wide**

FLOOR PLAN

### OPTIONS

*Sheet Plaster and Plaster Finish,* to take the place of wood lath, $146.00 extra. See page 109.
*Oriental Asphalt Shingles,* for roof, instead of wood shingles, $39.00 extra.
*Oak Doors, Trim and Floors* in living and dining room. *Maple Floors* in kitchen and bathroom, $126.00 extra.
*Storm Doors and Windows,* $51.00 extra.
*Screen Doors and Windows,* galvanized wire, $33.00 extra.

For prices of Plumbing, Heating, Wiring, Electric Fixtures and Shades see pages 130 and 131.

*For Our Easy Payment Plan See Page 144*

# The SUN ROOM
## The Year Around Porch

### What Our Prices Include

At the prices quoted we will furnish all the material to build these sun room porches, consisting of:

**Framing Lumber,** No. 1 Douglas Fir or Pacific Coast Hemlock;

**Ready-Cut Lath;**

**Siding,** Clear Cypress or Clear Red Cedar, Bevel, for No. P3213; Clear Cypress Drop Siding for No. P3214;

**Sheathing,** Douglas Fir or Pacific Coast Hemlock;

**Flooring,** Clear Douglas Fir or Pacific Coast Hemlock;

**Finishing Lumber;**

**Trim,** Beautiful Grain Douglas Fir or Yellow Pine;

**Windows,** Seven Pairs Modern French Windows; Sash to Open In Made of Clear California White Pine;

**40-Lb. Building Paper; Eaves Trough and Down Spout;**

**Stratford Design Hardware for French Windows,** Consisting of Hinges, Bolts and Fasteners. Furnished in either lemon brass or old copper finish. See page 132. Sufficient nails of all sizes required.

**Paint for Three Coats Outside;**

**Shellac and Varnish** for Inside Trim;

**Sun Room No. P3214,** Roofed With Best-of-All Roofing, Guaranteed for 16 years. We furnish enough material to lay the roofing with a 16-inch lap and also roof lap cement. This makes an absolutely tight roof;

**Sun Room No. P3213,** We Furnish Oriental Slate Surfaced Shingles, Guaranteed for 17 Years;

**Size of Sun Room Porch,** 10 feet by 12 feet;

**Height** from Floor to Ceiling, 9 feet;

**Foundation,** Concrete to Grade Line; Brick Wall Above.

THE modern sun room is a beautiful and useful addition to the home. Here dull days are spent pleasantly, a book may be read in solid comfort amidst flowers. Or to sit and listen to the radio music, entertain friends with assurance of a cheerful atmosphere. Others prefer it as a play room for their children. During the warm season, the sun room is the coolest spot in the house; during the winter it is warm and comfortable.

Illustrated are two popular sun room additions. These sun room porches are planned so that they can be added to a house already built at the lowest cost. Our price includes all the material for both inside and outside construction. The cost of a sun room is very little compared to the actual increased value of your home.

The Sun Room No. P3213 is the more suitable for one story or smaller homes. The Sun Room No. P3214 is more suitable for the larger or two-story residences. Select either sun room design and you may be sure of satisfaction.

**Sun Room Porch No. P3214**

## $329.00

For detailed description see upper left hand corner of this page.

**Sun Room Porch No. P3213**

## $279.00

For detailed description see upper left hand corner of this page.

### OPTIONS

*Sheet Plaster and Plaster Finish, to take the place of wood, lath and plaster, $14.00 extra. See page 109.*

*Storm Windows, $36.00 extra.*

*Screen Windows, galvanized wire, $15.00 extra.*

NOTE — Our prices do not include jamb, door, door trim or hardware for door between sun room and house to which it is attached. These items should correspond in design and size with doors in adjoining room. Write for our Millwork Catalog E543MH for designs and prices on these goods.

*For Our Easy Payment Plan See Page 144*

# *Goodwall* SHEET PLASTER

## *Build Your Home in Less Time and at Less Cost*

Goodwall Sheet Plaster sold with any "Honor Bilt" modern home. Look for "OPTION" at bottom of page on which house you plan to build is described. It quotes the cost of Sheet Plaster and Plaster Finish to take the place of wood lath.

**You can save time in building and move into your home earlier if you cover the walls with Goodwall Sheet Plaster.** In thousands of homes, in every section of the country, Goodwall Sheet Plaster is giving the greatest satisfaction. **It is a fire resisting gypsum rock composition plaster of even thickness between two sheets of heavy cardboard.** These cardboards are so saturated with the fire resisting gypsum composition that they make Goodwall Sheet Plaster far more fire resisting than regular lath and plaster. It comes

in sheets in convenient sizes and takes the place of regular lath and plaster at less cost and with less labor.

## Clean, Lasting and Easy to Apply

**Illustration shows Goodwall Sheet Plaster applied to studding.** It is given a coat of our specially prepared Hard Plaster and Top Coat Hard Plaster Finish. You will then have fire resisting walls, with a perfectly smooth surface, which can be tinted any desired color or decorated like any other plastered wall.

Actual Thickness ⅜-Inch Sheets

## Use Goodwall Sheet Plaster

*It enables you to do a better job for less money and can be applied by yourself.*

*It is better in service and strength than a lath and plaster wall.*

*It does away with lath stains.*

*It deadens sound as effectively as lath and plaster.*

*It is vermin proof and can be painted, papered or kalsomined.*

*It can be covered with a plaster coat of any thickness, but ⅜-inch Hard Plaster and ⅛-inch Top Coat Hard Plaster will be sufficient to make a smooth and seamless wall.*

*It is dry. No need to wait for weeks to drive the dampness out.*

*It can be removed in sections for plumbing repairs.*

*It is clean. No muss and dirt when you apply it.*

*It is strong. Send for a sample and try to break it with your hands.*

*It cuts out cost of lath; none needed with Goodwall Sheet Plaster.*

*It will not break when nailed. It is tough and durable.*

*It can be cut with a saw, or broken with a clean edge by scoring the cardboard on both sides with a knife, breaking where scored.*

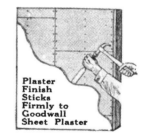

Plaster Finish Sticks Firmly to Goodwall Sheet Plaster

# Save $100 to $500 on Your Barn!

## By Our "Already Cut" System

Be sure to get our Barn Catalog E504MH, if you are planning to put up a barn or other farm buildings. Our Farm Buildings are furnished "Already Cut" in the same way as our Modern Homes. By this "Already Cut" System you can save nearly one-half of your carpenter labor. Moreover, you will get a stronger and better looking barn than by the old style of hand-cut barn building.

### $695.00
Size, 28x34 Already Cut and Fitted

The material is guaranteed to be the best of its kind and comes direct from our own big mills. This also gives you an additional saving.

Write for Your Free Copy Modern Farm Buildings Catalog E504MH

Our Farm Building Catalog contains a big variety of scientifically planned Barns, Hog Houses, Corn Cribs, Granaries, Implement Sheds, Poultry Houses and Silos.

No matter what kind of a farm building you are going to put up, we can save you money. Ask for your free copy of Farm Building Catalog, E504MH. Write now!

# High Grade "Honor Bilt"

We use only good building materials and honest construction. "Honor Bilt" homes reflect careful planning, even to the smallest detail. Read what our customers say throughout this catalog. We guarantee satisfaction.

No. 63MH391
*Entrance Door*

No. 63MH9861
*Mirror Door*

No. 63MH9865
*Built-In Medicine Case*

No. 63MH7009
*Casement Sash*

No. 63MH386
*Entrance Door*

No. 63MH7186
*Two-Light Window*

No. 63MH7015
*French Window*

No. 63MH7172
*Divided Light Window*

No. 63MH9850
*Entrance Door*

No. 63MH193
*Interior Door*

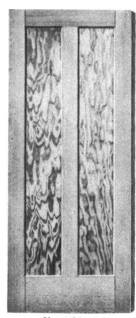

No. 63MH194
*Interior Door*

No. 63MH814
*Entrance Door*

No. 63MH191
*Interior Door*

# Millwork for Modern Homes

These built-in conveniences that will add so much to the family comforts are exclusively "Honor Bilt" Modern Home features. They are to be had in many of our homes in this book.

**No. 63MH9849**
*Colonial Brick Mantel*

**No. 63MH9851**
*Attractive Brick Mantel*

**No. 63MH9828**
*Triple Unit Clothes Closet*

We illustrate some of the high grade products furnished for our different Modern Homes, according to plans and specifications. The triple clothes closet, built-in features, doors, windows and frames are manufactured in our own millwork factory, one of the largest in the country. Expert mechanics, modern machinery and good materials insure perfectly made millwork.

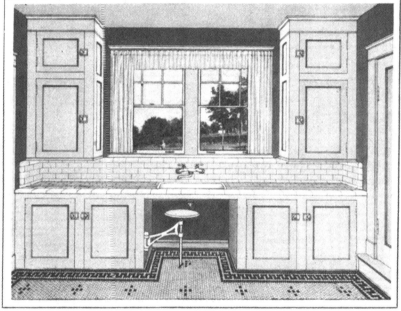

**No. 63MH9210**
*Kitchen De Luxe Outfit*

**No. 63MH9828**
Triple Unit Clothes Closet. The best closet arrangement known in architecture. Room for the most complete wardrobe in which garments are kept in better condition than in ordinary closets. Has hat or blanket compartment with three doors. There are three drawers at the bottom for shoes. No waste space. Three mirror doors, so hung that one can see one's self from all sides.

**No. 63MH9868**
*Built-In Ironing Board Outfit*

*No. 63MH9884 Upper Cupboard*
*No. 63MH9883 Lower Cupboard*

**No. 63MH9870**
*Breakfast Alcove*

# What Contractors Think of Our HonorBilt Ready Cut System

Sears, Roebuck and Co.
We have erected fourteen Sears, Roebuck and Co.'s "Ready Cut" houses. Your system saves actual cash for the builder. The materials furnished are the best and every piece of lumber is cut to fit and the detailed plans provided show just where and how each piece goes into the building.
RAYNER AND SNYDER

Sears, Roebuck and Co.
We are very well pleased with your "Ready Cut" Method and recommend your homes to anyone who desires a 100% house at a saving from $500.00 to $1,500.00. Your ready cut system fits perfectly and saves labor. Your materials are the best.
BLUME BROS.
Per Elmer M. Blume
Oscar W. Blume

Sears, Roebuck and Co.
Your "Honor Bilt" ready cut system of Modern Homes has my endorsement. I have built six of your houses, and I find your construction is easy to follow, takes less time to build, reduces costs, and eliminates waste; Your "Honor Bilt" Homes are permanent, high grade and satisfactory in every detail.
EMIL GOLKE

Sears, Roebuck and Co.
I am strong FOR THE READY CUT AND FITTED METHOD. Your guarantee of quality and quantity of materials relieves the contractor of this responsibility.
There is a very substantial saving in time in having everything cut to fit. The materials furnished are always good.
CLARENCE COLLISI

"HONOR BILT" Homes are not only the choice of thousands of families, but are also the choice of reliable and successful contractors everywhere.

"Honor Bilt" Homes are recommended by more than 34,000 owners, and by the independent contractors who build them for our customers.

Read what our customers say about "Honor Bilt" Homes on other pages in this book and the circular sent with it. On these two pages we reproduce just a few photographs and letters of reliable contractors who specialize in constructing "Honor Bilt" high grade modern homes.

These letters tell the whole story—vital facts that lead to but one answer—

## The Hallmark of "HONOR BILT" HOMES

High quality material throughout. "Honor Bilt" ready cut construction. Built for permanence, which means stronger and better homes without waste of material. Free architectural service. Built in less time at less cost. Built for comfort, warmth and beauty. A complete service without middlemen's profits. Guaranteed to satisfy.

Sears, Roebuck and Co.
The material used in your "Honor Bilt" Modern Homes is of excellent quality, the millwork particularly being a better quality than that which is ordinarily furnished by local dealers. Your ready cut materials are easily assembled, the plans and specifications are very clear.
P. E. REDDEN

Sears, Roebuck and Co.
The nineteen "Honor Bilt" Houses which I have erected have been a pleasure to handle. I have been proud of being the builder and to show home owners your materials and the way they come from your mills ready to assemble. Through your efficient service I have completed an average of a house every eighteen working days.
JAMES ENGLISH

Sears, Roebuck and Co.
I have constructed six of your houses. The material furnished and your method of construction is the best that I have ever used in my thirty-five years as a carpenter-contractor. Your method of ready cutting and detail plans work out exactly. I have always found ample material to complete each and every home.
WALTER SCHOBY

Sears, Roebuck and Co.
I have built twenty-two of your "Honor Bilt" Modern Homes and have always found your material to be high class, accurately fitted and in ample quantities to complete the jobs according to the plans and specifications. I am able to erect your houses with "handy men" instead of experienced carpenters.
WM. J. KING

Sears, Roebuck and Co.
I have built many of your "Honor Bilt" Homes. Your "Ready Cut" method of erecting homes makes for solid construction, saving of time and labor and money for the owner because you own and operate your own factories and millwork and lumber plants.
W. S. WAITE

Sears, Roebuck and Co.
I have always found your quality of the service very best, and price low, good and the "Ready Cut" plan of house construction is fundamentally sound, and effects a substantial saving in material and labor. My customers have always been pleased with the designs, practicability and convenience of your houses.
C. E. DAWSON

Sears, Roebuck and Co.
In 1916 I built the Gladstone, one of the first "Honor Bilt" Houses put up in Dayton, and have been using your material ever since. If people would just investigate the construction of these homes when planning their homes, they would all be building this type of construction and would be saving from $500 to $2,000.
CHAS. W. SHARP

Sears, Roebuck and Co.
I have signed the contract for my two hundred and fifty-seventh "HONOR BILT" Ready Cut House. This record could not have been made but for the fact of the genuine quality of the material; the practical advantages of the Ready Cut Method, and the saving.
IRVIN PAULSON

NOTE
Addresses of these and other contractors of "Honor Bilt" Homes furnished upon request.

# *Introducing—*

# STANDARD BUILT CONSTRUCTION *!*

**8 SPECIAL FEATURES** *of* **STANDARD BUILT LOW PRICED HOMES**

1. Good Quality Materials.
2. Ready-Cut Construction.
3. Modern Architectural Designs.
4. Most House per Dollar Invested.
5. No Extras: No Waste of Material.
6. Direct From Factory-to-You.
7. Save Nearly One-Half Labor Cost.
8. Machine Cut to Mathematical Correctness.

NOTE—For actual difference between Standard Built and "Honor Bilt" Houses see pages 12 and 13

*See Our Paint Colors on Page 20*

## All Standard Built Houses Are Sold for Cash or Easy Payments

### *Specifications for Standard Built Construction:*

**OUTSIDE WALL PLATES.** 2x6 No. 1 Douglas Fir or Pacific Coast Hemlock.

**CENTER GIRDER.** 6x8 No. 1 Douglas Fir or Pacific Coast Hemlock, built up of 3 pieces, 2x8, accurately cut to fit.

**BOX SILLS.** 2x8, No. 1 Douglas Fir or Pacific Coast Hemlock, cut to exact size.

**FLOOR JOIST.** 2x8 No. 1 Douglas Fir or Pacific Coast Hemlock, spaced 24 inches on center, accurately cut to fit.

**STUDDING.** 2x4 No. 1 Douglas Fir or Pacific Coast Hemlock, spaced 16 inches on center, cut to exact size for all walls and partitions.

**CEILING JOIST.** 2x4 No. 1 Douglas Fir or Pacific Coast Hemlock, spaced 16 inches on center, cut to exact size.

**RAFTERS.** 2x4 No. 1 Douglas Fir or Pacific Coast Hemlock, spaced 24 inches on center, accurately cut to fit so they can be set in place without any further cutting.

**PLATES.** 2x4 No. 1 Douglas Fir or Pacific Coast Hemlock, cut to exact size. A single plate is used at the bottom and a double plate is used at the top of all walls and partitions.

**ROOF SHEATHING.** 1x4 Douglas Fir or Pacific Coast Hemlock, spaced 2 inches apart to receive wood shingles. Sheathing for eaves, 1x6 dressed and matched. Furnished in random lengths.

**SHINGLES** for roof, Star A Star, 6-2 Red Cedar Shingles, to be laid 4½ inches to the weather.

**LATH** Douglas Fir 4-foot lath for all walls, partitions and ceilings.

**SIDING.** 1x6 No. 1 Douglas Fir or Pacific Coast Hemlock drop siding for all outside walls and gables.

**OUTSIDE FINISH.** Corner boards facia, frieze water table, moldings, etc., good sound well seasoned material, well suited for outside finish.

**PORCH FLOORING.** 1x4 dressed and matched edge grain Douglas Fir flooring.

**FINISH FLOORING.** 1x4 dressed and matched No. 1 Douglas Fir flooring.

**GROUND, BRIDGING, ETC.** Douglas Fir.

**DOOR AND WINDOW FRAMES** of select Douglas Fir with 1x4 outside casings, cypress sills for windows and oak sills for doors. Furnished knocked down and bundled.

**WINDOW SASH** of clear white pine, size 28x26, glazed with "A" quality glass firmly tacked and puttied.

**OUTSIDE DOORS.** 2 feet 8 inches by 6 feet 8 inches fir, with upper half glazed, mortised for locks.

**INSIDE DOORS.** Standard size five cross panel of fir, mortised for locks.

**INSIDE CASINGS.** Clear Douglas Fir or Yellow Pine; furnished in convenient lengths and bundled in sets.

**PORCH STEPS, RAILS AND BALUSTERS** of suitable material for the purpose.

**BASEBOARDS, PICTURE MOLDING AND INTERIOR TRIM.** Clear Douglas Fir or Yellow Pine.

**HARDWARE.** Lemon brass finish mortised lock sets, door butts, sash locks and sash lifts. Sash weights, sash cord and tarred felt.

**NAILS.** Sufficient quantity of various sizes of nails required for all work.

**PAINT.** Sufficient quantity of paint with necessary oil for thinning for two coats for outside body and trim.

**SHELLAC AND VARNISH** for interior doors and trim of sufficient quantity for two-coat work.

*PLANS—Complete Working Plans Showing Clearly How Your Standard Built Ready-Cut House Is to Be Constructed*

# THE GRANT
## Standard Built
### No. P6018—Already Cut and Fitted
## $999.00

THE GRANT Standard Built Home is an example of down-right value. This is made possible by our splendid manufacturing organization and by our ability to buy raw material in large quantities at favorable prices. While the Grant is not built of the same high grade material as our "Honor Bilt" Homes, it is of better quality than similar houses offered elsewhere at higher prices.

The Grant is a practical home. Economy of space has been the watchword in its making, yet every one of its six rooms is large enough for the usual needs of a family.

**The Living Room.** You enter into the living room from the attractive front porch. Size, 13 feet 2 inches by 12 feet 2 inches. It has ample wall space for a piano and floor space for other furniture. Light and air provided by one front window and double side window.

**The Dining Room.** A cased opening connects both living room and dining room. Size, 13 feet 2 inches by 12 feet 1 inch. There is space for a dining suite. A double window provides light and ventilation.

**The Kitchen.** From the dining room a door connects with the kitchen. Space is provided for the sink, the range, the table and the ice box. One window provides light

and air. A rear door opens into the yard.

**The Bedrooms.** The front bedroom opens from the living room. It has a clothes closet. Light and cross current of air from one front window and one side window. The middle bedroom is entered from the dining room. Here, too, is a closet for clothes. Light and air from one window. The third bedroom is entered from an open hall off the dining room. One window provides light. The bathroom is also entered from this hall.

**Height of Ceiling.** 8 feet 3 inches from floor to ceiling.

**At the price quoted we will furnish all the material to build this six-room house and bath, consisting of lumber, lath, millwork, flooring, porch ceiling, shingles, siding, finishing lumber, tarred felt, hardware and paint material. We guarantee enough material to build this house. Price does not include cement, brick or plaster.**

## OPTIONS

*Storm Doors and Windows, $45.00 extra.*

*Screen Doors and Windows, galvanized wire, $34.00 extra.*

*Sheet Plaster and Finish, to take the place of wood lath, $138.00 extra. See page 109.*

*Oriental Asphalt Shingles, guaranteed 17 years, instead of wood shingles, for roofs, $60.00 extra.*

*For Our Easy Payment Plan See Page 144*

THE SELBY Standard Built Home has been designed with the same care as the more expensive houses. Its simple architecture has a universal appeal. Then, again, its four rooms really serve every purpose of five!

You benefit by our ready cut construction—costs less to erect this home.

No middlemen's profits. Consider our low price for such a good little home!

**The Living and Dining Room.** The front porch leads to the combination living and dining room. Size, 15 feet 8 inches by 9 feet 8 inches. There is plenty of wall space for a piano, furniture and dining room set. Three front windows provide plenty of light and air.

**The Kitchen.** A door connects the living and dining room with the square kitchen. Size, 9 feet 2 inches by 9 feet 2 inches. Wall space accommodates the sink, range, kitchen cabinet, table and ice box. A window affords light and air. A glazed door leads to the yard.

**The Bedrooms.** The front bedroom opens directly off the living and dining room, and has two windows, providing light and cross current of air.

# THE SELBY

## Standard Built

### No. P6011—Already Cut and Fitted

# $629.00

### A Living and Dining Room, Kitchen and Two Bedrooms With Bathroom

An open hallway off the living and dining room connects with the rear bedroom. One window provides light and air.

The hallway has a roomy clothes closet and a door leading to the bathroom, which has a window.

**Height of Ceiling.** 8 feet 3 inches from floor to ceiling.

*For Our Easy Payment Plan See Page 144*

FLOOR PLAN

At the price quoted, we will furnish all material to build this four-room house and bath, consisting of lumber, lath, millwork, flooring, shingles, siding, finishing lumber, tarred felt, hardware and paint material. We guarantee enough material to build this house. Price does not include cement, brick or plaster.

## OPTIONS

*Storm Doors and Windows, $35.00 extra.*

*Screen Doors and Windows, galvanized wire, $27.00 extra.*

*Sheet Plaster and Finish, to take the place of wood lath, $88.00 extra. See page 109.*

*Oriental Asphalt Shingles, guaranteed 17 years, instead of wood shingles for roofs, $27.00 extra.*

THE KIMBALL Standard Built is a sensible four-room home. It offers an opportunity to thousands of rent payers to own a home of their own at a modest outlay of money.

It is made of good material, planned so that every foot of lumber is used to the best advantage, with the rooms laid out in a pleasing way.

**The Living Room.** You enter the living room from the porch. Size, 13 feet 2 inches by 11 feet 8 inches. A long wall space is provided for a piano. There is room for a davenport and other furniture. One front window and a side window provide light and cross ventilation.

**The Kitchen.** A door connects the living room and kitchen. Size, 13 feet 2 inches by 11 feet 2 inches. Every need of a well laid out kitchen has been considered. There is room for the sink, range, pantry, ice box and dining table and chairs. A door leads to the yard.

**The Bedrooms.** Front bedroom is entered from the living room. Size of front bedroom, 9 feet 8 inches by 11 feet 8 inches. The rear bedroom opens from the kitchen.

# THE KIMBALL
### Standard Built

*No. P6015—Already Cut and Fitted*

## $635.00

**Our From Factory to You System Makes This Value Possible**

*A Living and Dining Room, Kitchen and Two Bedrooms*

Size of rear bedroom, 9 feet 8 inches by 11 feet 2 inches. Each bedroom has two windows.

**Height of Ceiling.** 8 feet 3 inches from floor to ceiling.

At the price quoted, we will furnish all material to build this four-room house, consisting of lumber, lath, millwork, flooring, shingles, siding, lattice, finishing lumber, tarred felt, hardware and painting material. We guarantee enough material to build this house. Price does not include cement, brick, or plaster.

FLOOR PLAN

### OPTIONS

*Storm Doors and Windows, $35.00 extra.*

*Screen Doors and Windows, galvanized wire, $26.00 extra.*

*Down Spout and Eaves Trough, $16.00 extra.*

*Sheet Plaster and Plaster Finish, to take the place of wood lath, $87.00 extra. See page 109.*

*Oriental Asphalt Shingles, guaranteed 17 years, instead of wood shingles for roofs, $39.00 extra.*

# THE RAMSAY

**Standard Built**

*No. P6012—Already Cut and Fitted*

# $685.00

*A Living and Dining Room, Kitchen, Two Bedrooms, With Bathroom*

THE RAMSAY Standard Built Home was designed with the aim of meeting a demand for a modest home at an equally modest price. It has a special appeal to those who live in suburban centers, who have a small family, or who want to live modestly in order to save a sum of money for a more elaborate home in the future.

**The Living and Dining Room.** The front door opens from the porch. Size of the combination living and dining room, 9 feet 8 inches by 15 feet 8 inches. The living room furniture is usually arranged towards the front while the dining suite is placed near the kitchen. There is a good wall space for a piano. One front window and two side windows provide plenty of light and air.

**The Kitchen.** A door opens from the living and dining room into the kitchen. It has space for sink, range, kitchen cabinet, and a table to be placed underneath the window. A door leads to the rear yard.

**The Bedrooms.** The front bedroom opens from the living and dining room. It has a clothes closet. One front window provides light and ventilation. The rear bedroom connects with the kitchen. A side window affords light and air.

**The Bathroom** connects with the living and dining room.

**Height of Ceiling.** Main floor, 8 feet 3 inches from floor to ceiling.

*For Our Easy Payment Plan See Page 144*

FLOOR PLAN

At the price quoted, we will furnish all material to build this four-room house with bath, consisting of lumber, lath, millwork, flooring, porch ceiling, shingles, siding, finishing lumber, tarred felt, hardware, and painting material. We guarantee enough material to build this house. Price does not include cement, brick, or plaster.

### OPTIONS

*Storm Doors and Windows, $32.00 extra.*
*Screen Doors and Windows, galvanized wire, $25.00 extra.*

*Sheet Plaster and Plaster Finish, to take the place of wood lath, $38.00 extra. See page 109.*

*Oriental Asphalt Shingles, guaranteed 17 years, instead of wood shingles for roofs, $37.00 extra.*

THE HUDSON Standard Built Home is an investment that readily appeals to the thrifty family. Here is everything one expects in a modest and up to date house. Consider the good quality material and construction and low price! The Hudson was planned by us after many comparisons of designs. All the material comes from our own large mills and yards. No middleman's profits.

**The Exterior** presents a neat and becoming appearance. The porch is sheltered underneath the main roof of the house.

**The Living Room.** From a porch a door opens into the living room. Size, 9 feet 5 inches by 10 feet 7 inches. There's ample wall space for a piano and furniture. A side window provides light and ventilation.

**The Kitchen.** From the living room a door opens into the kitchen. Size, 9 feet 5 inches by 9 feet 5 inches. It has plenty of space for sink, stove, kitchen table and ice box. Light and air from one window. A door leads to the rear yard.

**The Bedrooms.** The front bedroom is entered from the living room. A front window supplies light and air.

The rear bedroom is entered from the kitchen. One window supplies light and air.

**The Bathroom** is entered from the living room. Plumbing fixtures may be roughed-in on one wall, saving on installation.

**Height of Ceiling.** 8 feet 3 inches from floor to ceiling.

# THE HUDSON
### Standard Built
**No. P6013A—"Hudson" 4 Rooms and Bath**

## $499⁰⁰

At the price quoted we will furnish all the material to build this four-room house with bath, No. P6013A, consisting of lumber, lath, flooring, porch ceiling, siding, finishing lumber, millwork, slate surfaced roofing, hardware and paint. We guarantee enough material to build this house. Price does not include cement, brick or plaster.

**No. P6013—Already Cut and Fitted**

## $659⁰⁰

At the price quoted we will furnish all material to build this four-room house with bath, No. P6013, consisting of lumber, lath, millwork, flooring, porch ceiling, shingles, siding, finishing lumber, tarred felt, hardware and painting material. We guarantee enough material to build this house. Price does not include cement, brick or plaster.

### OPTIONS

*Storm Doors and Windows, $32.00 extra for No. P6013; $26.00 extra for No. P6013A.*

*Screen Doors and Windows, galvanized wire, $25.00 extra for No. P6013; $21.00 extra for No. P6013A.*

*Front and Rear Steps for No. P6013A, $9.00 extra.*

*Wood Shingles, instead of slate surfaced roofing, for No. P6013A, $22.00 extra.*

*Tarred Felt, under floor and siding, for No. P6013A, $13.00 extra.*

*Oriental Asphalt Shingles, guaranteed 17 years, instead of wood shingles, for roofs, $35.00 extra for P6013.*

*Sheet Plaster and Finish, to take the place of wood lath, $80.00 extra for P6013A. See page 109.*

*Sheet Plaster and Finish, to take the place of wood lath, $96.00 extra for P6013. See page 109.*

FLOOR PLAN No. P6013A

FLOOR PLAN No. P6013

# THE FARNUM

## Standard Built

*No. P6017—Already Cut and Fitted*

# $942.00

*A Living Room, Dining Room, Kitchen, Two Bedrooms, With Bathroom*

THE FARNUM Standard Built Home is patterned after the more expensive bungalow. It is well balanced in both exterior and interior. The material is of a good grade; in fact, better than is usually found in this kind of a house. The Farnum is priced in keeping with our policy of greater value giving. Our tremendous business, plus our direct-from-factory-to-you system, makes this possible.

**The Living Room.** From the front porch, which is 24 feet by 8 feet, one enters the living room through a handsome door. Size, 12 feet 8 inches by 12 feet 2 inches. It has wall space for a piano, and room for furniture. Two windows afford light.

**The Dining Room.** A cased opening connects the living room and dining room. Size, 12 feet 8 inches by 10 feet 1 inch. A double window throws light.

**The Kitchen.** Size, 10 feet 8 inches by 8 feet 2 inches. Space is provided for a sink, range, table and ice box. It has one window. A door leads to the rear yard.

**The Bedrooms.** An open hall off the dining room connects with the two bedrooms. Each bedroom has two windows, assuring light and ventilation. There is a clothes closet in each bedroom.

**The Bathroom** plumbing is to be roughed-in on one wall, saving part of installation expense.

**Height of Ceiling.** 8 feet 3 inches from floor to ceiling.

*For Our Easy Payment Plan See Page 144*

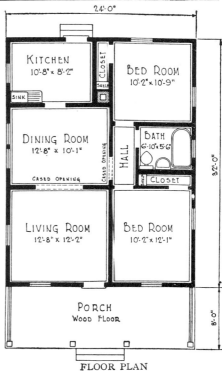

FLOOR PLAN

At the price quoted we will furnish all material to build this five-room house and bath, consisting of lumber, lath, millwork, flooring, porch ceiling, shingles, siding, finishing lumber, tarred felt, hardware and painting material. We guarantee enough material to build this house. Price does not include cement, brick or plaster.

### OPTIONS

*Storm Doors and Windows, $41.00 extra.*
*Screen Doors and Windows, galvanized wire, $32.00 extra.*
*Sheet Plaster and Plaster Finish, to take the place of wood lath, $128.00 extra. See page 109.*
*Oriental Asphalt Shingles, guaranteed 17 years, instead of wood shingles, for roofs, $57.00 extra.*

# THE ESTES

## Standard Built

### No. P6014—Already Cut and Fitted

## $672.00

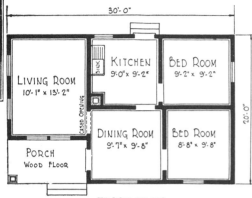

### FLOOR PLAN

THE ESTES Standard Built bungalow is an inviting little home, priced unusually low. Its front porch is entirely underneath the roof of the house and may be enclosed in glass and used as a sun room.

**The Dining Room** is entered from the porch. Size, 9 feet 7 inches by 9 feet 8 inches. A window provides light and air.

**The Living Room.** A cased opening to the left connects the dining room and the living room. Size, 10 feet 1 inch by 13 feet 2 inches. A piano and the furniture can be grouped attractively because of the large wall spaces. A double window overlooks the porch, and two windows are at the side, thus assuring an abundance of light and air.

**The Kitchen** is entered from the dining room. The sink, range, kitchen cabinet and table space are planned to save steps and time. Light and air are provided by a window. A rear door leads to the yard.

**The Bedrooms.** The front bedroom connects with the dining room. The other bedroom opens off the kitchen. Each bedroom has a window.

**Height of Ceiling.** 8 feet 3 inches from floor to ceiling.

At the price quoted, we will furnish all material to build this five-room house, consisting of lumber, lath, millwork, flooring, porch ceiling, shingles, siding, finishing lumber, tarred felt, hardware and painting material. We guarantee enough material to build this house. Price does not include cement, brick or plaster.

### OPTIONS

*Storm Doors and Windows, $36.00 extra.*

*Screen Doors and Windows, galvanized wire, $28.00 extra.*

*Sheet Plaster and Plaster Finish, to take the place of wood lath, $87.00 extra. See page 109.*

*Oriental Asphalt Shingles, guaranteed 17 years, instead of wood shingles for roof, $36.00 extra.*

---

# THE FOSGATE
## Standard Built
### No. P6016—Already Cut and Fitted
## $681.00

### FLOOR PLAN

### OPTIONS

*Storm Doors and Windows, $32.00 extra.*

*Screen Doors and Windows, galvanized wire, $25.00 extra.*

*Eaves Trough and Down Spout, $10.00 extra.*

*Sheet Plaster and Plaster Finish, to take the place of wood lath, $100.00 extra. See page 109.*

*Oriental Asphalt Shingles, guaranteed 17 years, instead of wood shingles for roof, $37.00 extra.*

THE FOSGATE Standard Built Home readily meets with favor because: It has an attractive exterior and a modern interior. It is made of good quality material, ready cut, and square. Priced on the bedrock of greatest value.

**The Living and Dining Room.** The front door opens directly from the porch into the combination living and dining room. Size, 14 feet 2 inches by 12 feet 8 inches. Wall space permits a piano and furniture, while the dining table and chairs can be placed near the kitchen side of the room. One front window and a double side window flood the room with the brightness of outdoors.

**The Kitchen.** A door connects the living and dining room with the kitchen. Size, 8 feet 2 inches by 10 feet 2 inches. Everything, including the sink, range, table, ice box, etc., can be placed to reduce

steps and time for the housewife. A rear door leads to the outside.

**The Bedrooms.** The front bedroom opens from the living room. The rear bedroom connects with the open hall off the living room. Each bedroom has a clothes closet, and a window.

**The Bathroom** also opens off the hall. Plumbing can be roughed-in on one wall, thereby lowering construction cost.

**Height of Ceiling.** 8 feet 3 inches from floor to ceiling.

At the price quoted, we will furnish all material to build this four-room house and bath, consisting of lumber, lath, millwork, flooring, shingles, siding, finishing lumber, tarred felt, hardware and paint material. We guarantee enough material to build this house. Price does not include cement, brick or plaster.

# *Simplex Sectional*
# GARAGES and SUMMER COTTAGES
## Come in sections quickly bolted together; no sawing, no nailing; easily handled

**Do not confuse the construction of these buildings** with our "Honor Bilt" Houses fully described on pages 1 to 112, or "Standard Built" houses on pages 113 to 120. Simplex Sectional Buildings come in sections (see illustrations below). They can be erected in two to three days; garages in a much shorter time, all carpenter labor having

*Eight o'Clock*

The first picture, taken a moment before eight o'clock, shows the sections of Sectional Garage 55P22 on the ground ready for erection.

*Half Past Nine*

One hour and thirty minutes later, with the garage well under way. Sills have been bolted at the corners, squared and leveled. The wall sections have been assembled and the triangular gable pieces of the roof connected by ridge pole and roof supports. Two of the roof sections are in position.

*One o'Clock*

By one o'clock the main part of the building has been completed. The doors have been hung, all of the roof sections in place and part of the roofing has been laid.

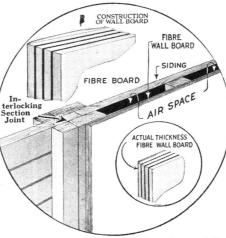

*Quarter Past Two*

The last picture, taken after five hours and forty-five minutes of work, shows the garage completed ready for use. The roofing has been applied and the ridge roll has been fastened along peak.

*Illustration above shows how by applying wall board to the interior of your sectional cottage you get double wall construction. Wall board quoted as an option for each frame cottage.*

been done at the factory. Note the illustrations at left below which are actual photographs of the erection of one of our garages.

Simplex Sectional Buildings make ideal summer cottages and substantial garages. They are made by expert mechanics in our own electrically driven factory, especially designed, equipped and manned for this particular kind of work. These men devote their entire time to this work, and you are assured of neat, accurate and durable craftsmanship. The material used in these buildings is carefully selected for strength, wear and appearance.

## Specifications for Simplex Sectional Cottages and Garages

**FOUNDATION**—Cedar posts, 6 inches in diameter and 4 feet long, are furnished for cottages. Garages are often placed on concrete foundations, so no posts, floor joists, or flooring are furnished for them. Two-foot posts for garages shown as option, at small extra price.

**SILLS**—3x6 No. 1 Oregon Fir or Pacific Coast Hemlock, which rest on foundation or posts.

**FLOOR JOISTS**—For cottages, 2x6 No. 1 Oregon Fir or Pacific Coast Hemlock, hung in patent steel stirrups 16 inches on center.

**FLOORING**—For cottages, No. 1 Oregon Fir or Pacific Coast Hemlock, 3¼-inch face. Floors shipped in sections 3 feet wide. Garages are not furnished with joists or floors.

**WALL SECTIONS**—No. 1 Oregon Fir or Pacific Coast Hemlock siding 1 inch thick, finished standard with tongue and groove, smooth on inside. Sections, 3 feet wide and 8 feet high. Sections tongue and groove together with reinforced interlocking joint, made of three thicknesses of one-inch lumber. See illustration in circle at bottom of page.

**INSIDE PARTITIONS**—No. 1 Oregon Fir or Pacific Coast Hemlock beaded ceiling ⅝ inch thick. Sections, 3 feet wide, 7 feet 10 inches high.

**ROOF SECTIONS**—Three feet wide No. 1 Oregon Fir or Pacific Coast Hemlock beaded ceiling, ⅝ inch thick.

**ROOFING**—Garages furnished with sea green Oriental Slate Surfaced Roofing; cottages with Fire-Chief Shingle Roll Roofing, both guaranteed for 17 years. Roofing shipped in rolls with nails and cement. Red roofing will be furnished instead of green on request.

**DOORS AND WINDOWS**—Doors are 2 feet 6 inches by 6 feet 6 inches, 1⅜ inches thick; complete with hardware, including fitted mortised locks. Inside doors, five-cross panel. Outside doors, upper half glazed clear glass. French doors, where specified, are 4 feet wide by 6 feet 6 inches high, ten lights of glass in each door. All windows, two-light glazed, size 24x26 inches, hung complete with spring bolt locks. French windows, where specified, are 2 feet 4⅛ inches by 4 feet 10 inches, ten lights of glass in each sash.

**Our Famous Garage Doors** equipped with special patented hardware, fully described on page 124, are furnished with Simplex Sectional Wood Garages only. Doors are 8 feet by 8 feet—substantial—strong and conform to illustration shown on each garage.

**PAINT**—The entire outside of cottages and garages painted one coat of gray trimmed in white.

**SCREENS**—Furnished in 14-mesh galvanized wire screen when ordered.

**SCREENS FOR PORCHES**—When ordered, we omit porch railing, balusters and columns, and ship regular screen sections covered with 14-mesh galvanized wire netting. (See illustration below.)

**WALL BOARD**—Sectional cottages can be lined with wall board, forming a double wall with air space between. See illustration in circle below. Wall board is cut to proper sizes, and shipped complete for walls and ceiling with necessary lumber accessories.

**SHIPMENT** of frame sectional buildings made complete from factory in Newark, N. J. Steel garages shipped from factory in Ohio.

**TERMS**—Sectional summer cottages sold for cash only. Garages sold for cash or on our LIBERAL EASY PAYMENT PLAN. See order blank on page 129.

**PLANS**—Complete instructions and plans explaining simple method of assembling furnished with each building.

**MASONRY** items, steps and plumbing are not included in the price of building.

**Gable Section**

**Outside Door Section**     **Inside Door Section**     **Colonial Window Section**

**Screened Porch Section**     **Inside Partition Section**     **Outside Wall Section**

# *Simplex Sectional* GARACES

## *The* AVENUE

### 55P32 Frame Garage

## $173.00

The Avenue is a strong, well built hip roof garage, which will harmonize well with any style of architecture. It is furnished in six sizes for one or two cars, and can be furnished for more than two cars for a small additional price for each additional 9-foot section. This garage is sold for cash or easy payment terms. See our order blank on page 129. The single car garage is furnished with three windows, one service door, and Sears triple sliding glazed auto doors. The two-car garages have two sets of sliding doors.

| Catalog No. | Foundation Dimensions | | | | Price | Shipping Weight, Pounds | Add for Foundation Posts |
|---|---|---|---|---|---|---|---|
| | Ft. | In. | Ft. | In. | | | |
| 55P32 | 12 | 3 | x 15 | 3 | $173.00 | 3,600 | $2.00 |
| 55P33 | 12 | 3 | x 18 | 3 | 190.00 | 4,400 | 2.00 |
| 55P34 | 12 | 3 | x 21 | 3 | 204.00 | 4,900 | 2.25 |
| 55P37 | 18 | 3 | x 18 | 3 | 256.00 | 5,300 | 2.25 |
| 55P38 | 21 | 3 | x 18 | 3 | 279.00 | 6,400 | 2.50 |
| 55P39 | 21 | 3 | x 21 | 3 | 299.00 | 7,200 | 2.75 |

*These garages are equipped with Sears Triple Sliding Doors. For complete description of these doors see page 124*

Plan of Double Garage

DOTTED LINES SHOW POSITION OF TRIPLE SLIDING DOORS WHEN OPEN.

## *The* MANOR

### 55P43 Frame Garage

## $227.00

This is our most attractive garage, being constructed with a double wall. The inside wall is tongue and groove lumber. The outside wall is paneled with Oriental Slate Surfaced Siding as shown in the illustration, giving a very artistic stucco effect. The sizes are unusually large, allowing plenty of room to work around the car. These garages are furnished with five windows and a service door and Sears triple sliding glazed doors. The two-car garages (18 feet 3 inches and 21 feet 3 inches wide) contain two sets of Sears triple sliding glazed doors. This garage can be furnished for more than two cars at a small additional cost for each 9-foot section.

|←——15'-3"——→|

GARAGE

Furnished in Five Sizes.

DOTTED LINES SHOW POSITION OF TRIPLE SLIDING DOORS WHEN OPEN.

| Catalog No. | Foundation Dimensions | | | | Price | Shipping Weight, Pounds | Add for Foundation Posts |
|---|---|---|---|---|---|---|---|
| | Ft. | In. | Ft. | In. | | | |
| 55P43 | 12 | 3 | x 18 | 3 | $227.00 | 4,700 | $2.00 |
| 55P45 | 15 | 3 | x 18 | 3 | 239.00 | 5,200 | 2.00 |
| 55P46 | 15 | 3 | x 21 | 3 | 258.00 | 6,100 | 2.25 |
| 55P48 | 21 | 3 | x 18 | 3 | 306.00 | 6,200 | 2.50 |
| 55P49 | 21 | 3 | x 21 | 3 | 329.00 | 7,200 | 2.75 |

*These garages are equipped with Sears Triple Sliding Doors. Read what wonderful doors these are. See page 124*

**For General Specifications See Page 121**

# *Simplex Sectional* GARAGES

## The SECURITY

### 55P10 Frame Garage

## $132.00

If you have a limited space and wish to put up a garage you will find that the single car Security garage is just what you want. It is furnished with our famous Sears Triple Sliding Doors, one of which can be used as a service door. The single car garage is furnished with one sash, 28x30 inches, without side door. Two-car garage 18 feet 3 inches wide has two sets of sliding doors as shown in the illustration, two sash and one side door, 2 feet 6 inches by 6 feet 6 inches. Sold for cash or on liberal easy payment terms. You cannot afford to pay rent on a garage when you can own one so easily.

GARAGE
Furnished In Four Sizes

Dotted Lines Show
Position of Triple
Sliding Doors When Open

| Catalog No. | Foundation Dimensions | | | | Price | Shipping Weight, Pounds | Add for Foundation Posts |
|---|---|---|---|---|---|---|---|
| | Ft. | In. | | Ft. | In. | | |
| 55P10 | 9 | 3 | x | 15 | 3 | $132.00 | 2,500 | $1.50 |
| 55P11 | 9 | 3 | x | 18 | 3 | 148.00 | 2,800 | 1.50 |
| 55P13 | 12 | 3 | x | 18 | 3 | 168.00 | 4,300 | 2.00 |
| 55P17 | 18 | 3 | x | 18 | 3 | 247.00 | 5,200 | 2.25 |

*These garages are equipped with Sears Triple Sliding Doors. Read what wonderful doors these are. See page 124*

## The PARKWAY

### 55P22 Frame Garage

## $162.00

The Parkway is an artistic and roomy garage which can be bought for cash or easy payments. Our easy payment plan enables you to invest the money, wasted on rent, in your own garage. The investment will pay for itself many times over.

The illustration shows a single car garage which is furnished with glazed garage doors and two full size glazed windows. The two-car garages (18 feet 3 inches and 21 feet 3 inches wide) have two sets of Sears Triple Sliding Glazed Doors and a service door, 2 feet 6 inches by 6 feet 6 inches.

GARAGE
Furnished in Six Sizes

Dotted Lines Show Position
of Triple Sliding Doors
When Open

| Catalog No. | Foundation Dimensions | | | | | Price | Shipping Weight, Pounds | Add for Foundation Posts |
|---|---|---|---|---|---|---|---|---|
| | Ft. | In. | | Ft. | In. | | | |
| 55P22 | 12 | 3 | x | 15 | 3 | $162.00 | 3,500 | $2.00 |
| 55P23 | 12 | 3 | x | 18 | 3 | 175.00 | 4,300 | 2.25 |
| 55P24 | 12 | 3 | x | 21 | 3 | 192.00 | 4,800 | 2.25 |
| 55P27 | 18 | 3 | x | 18 | 3 | 255.00 | 5,200 | 2.25 |
| 55P28 | 21 | 3 | x | 18 | 3 | 279.00 | 6,400 | 2.50 |
| 55P29 | 21 | 3 | x | 21 | 3 | 299.00 | 7,200 | 2.75 |

*These garages are equipped with Sears Triple Sliding Doors. For complete description of these doors see page 124*

**For General Specifications See Page 121**

# THE PERFECT GARAGE DOOR

## Works Easy ~
## Never gets out of order

Progress comes to the aid of twenty million motor car owners. It means the end of banging and slamming garage doors hung on hinges. The old kind of door is no longer used when the Sears Triple Sliding Doors are to be had. The old fashioned hinged door broke many automobile lamps and fenders when a gust of wind would send the heavy, cumbersome door against the car while entering the garage.

The unique Patented Roller Bearing Sliding Door Equipment and doors are almost worth the price of the garage alone, because they will pay for themselves in a short time, if the cost of damages to the car is considered. The Sears Garage Door cannot close at the critical moment. Neither do you have to shovel snow and ice away in the winter time in order to open it.

The Sears Garage Door moves easily and quickly over a steel track, on roller bearing rollers, balanced perfectly. A ten-year old child can open the doors without exertion. The rollers cannot get off the track.

*No Other Garage Doors Compare With the Sears Triple Sliding Door*

See illustration "A" above which shows how the doors slide close to the wall with no loss of space. They take up no room except 13 inches of waste space in one corner.

One door serves two purposes: One door is hinged, to be used as a service door as well as a sliding door.

The Sears Triple Sliding Door Equipment costs over three times as much as the ordinary kind. Yet, we offer it to you as part of our Simplex Garages at NO ADDITIONAL COST.

You will be more than satisfied if you purchase a Simplex Sectional Garage, equipped with the Sears Triple Sliding Door.

---

### Simplex Portable Outhouse
### 4 Ft. Square

## $41.00

**55P14**
Shipping weight, 720 pounds.

Wall sections, 4 feet wide and 7 feet high to eaves. Side ventilator opening, diamond shape, on each side. Contains two comfortable, well finished seats, one for adults, one for children. This is a neat, attractive, little building. Best-of-all Asphalt Roofing. Color, black.

## Hunters' Cabin or Trap Shooting House

## $238.00

**55P1**—One-room sectional building, 12 x 9 feet. Shipping weight, 4,800 pounds.

The ease with which this building can be erected appeals to the hunter, who wants a cabin in a hurry; one that he can put up in a day or two, and one that can be moved to a new location as the hunting grounds change. It is a one-room building 12 feet wide by 9 feet deep with plenty of windows and a large porch 8 feet deep, 12 feet wide. The solid door in the front and batten shutters as shown in the illustration for the windows permit the building to be closed up tightly when not in use. These are furnished in the price shown above. Roofing: Best-of-all black asphalt roofing, guaranteed for 14 years, is furnished for the cabin.

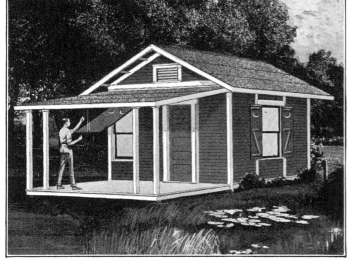

# Special Sectional GARAGES

While the garages shown on this page are not Simplex construction, they are strong and will give the service that you have a right to expect. The framework consists of 2x2's and 2x4's, which form part of the wall sections, so that the whole building can be easily fastened together with bolts and screws which we furnish. Garages contain hinged swinging doors.

WALL SECTIONS—Select No. 1 Oregon Fir or Pacific Coast Hemlock, tongued and grooved drop siding 1 inch thick, finished standard.
DOUBLE DOORS—No. 1 Oregon Fir or Pacific Coast Hemlock, ¾-inch matched and beaded partition. Opening size, 8 feet wide. Strongly built. Fitted and ready to hang with three pairs of large steel strap hinges.
WINDOW—One four-light glazed sash hinged at the top. 28x30 inches.
ROOFING—Best-of-all black asphalt, 50-pound rolls, with nails and cement.
PAINT—Painted one coat of gray trimmed in white.

## The ECONOMY

### 55P50 Frame Garage

## $86.50

"Economy" garages are flat roof garages, popular in many sections of the country. They are primarily designed as sturdy and neat garages at the lowest possible prices. They come in sections easily bolted together by any handy man. You can erect it yourself over the week end, and save the money you would pay some one else to build it for you. The price is so low that less than a year's rent will pay for it. You cannot afford to rent a garage when you can own one so easily.

These garages are 8 feet high in front, sloping to 7 feet 6 inches in rear. The roof is made of 1x6-inch No. 2 dressed and matched, covered with our Best-of-all Black Asphalt Roofing, guaranteed for 14 years. The 18x18-foot garage pictured is a two-car garage furnished with two sets of swinging doors.

| Catalog No. | Foundation Dimensions | Price | Shipping Weight, Pounds | Add for Foundation Posts |
|---|---|---|---|---|
| 55P50 | 10 x 14 | $ 86.50 | 2,400 | $1.25 |
| 55P51 | 10 x 16 | 93.00 | 3,000 | 2.00 |
| 55P52 | 10 x 18 | 101.00 | 3,600 | 2.00 |
| 55P57 | 18 x 18 | 153.00 | 4,800 | 2.52 |

## The UTILITY

### 55P60 Frame Garage

## $99.00

"The Utility" garage has a gable roof adding to its attractiveness. Like "The Economy" it is a substantial building and an excellent value at the price offered. Our price is low and you are sure to make a big saving by buying this garage. It is also sold on the Easy Payment Plan, which will make it possible for you to own a garage without paying more than rent. It can be erected by any handy man. Put it up yourself over the week end. "The Utility" is furnished with 2x6-foot service door, making it unnecessary to open the large doors each time you want to go in. The roof sections are made of ⅝x4-inch Select No. 1 Oregon Fir or Pacific Coast Hemlock, dressed and beaded on the under side, covered with our Best-of-all Roofing, guaranteed for 14 years. The 18x18-foot garage is a two-car garage furnished with two sets of swinging doors.

| Catalog No. | Foundation Dimensions | Cash Price | Shipping Weight, Pounds | Add for Foundation Posts |
|---|---|---|---|---|
| 55P60 | 10 x 14 | $ 99.00 | 2,600 | $1.25 |
| 55P61 | 10 x 16 | 109.00 | 3,200 | 2.00 |
| 55P62 | 10 x 18 | 117.00 | 3,800 | 2.00 |
| 55F67 | 18 x 18 | 170.00 | 5,000 | 2.25 |

# SECTIONAL STEEL GARAGES

GARAGES on this page are all steel, made of new "open hearth" 26-gauge heavily galvanized steel sheets. They are fireproof, stormproof, substantial and very simple in construction. The interchangeable parts can be easily put together by any handy man. No soldering, riveting or machine work of any kind is necessary. They are bolted together. In each corner two 14-gauge solid steel braces are furnished to make the building very rigid. These steel braces also form very convenient shelves.

## The MONITOR — $129.00 And Up

### 55P70 Sectional Fireproof Steel Garage

By using a curved corrugated roof on this garage, it greatly increases the strength of this roof over the usual type of steel garages. This shape of roof also adds much to the appearance of the building, making it very attractive. If windows are required in the side or rear, they can be furnished for $6.50 each, including steel frame hinged at the top and sash glazed with reinforced wire glass, 16x20 inches. Side door, 2 feet by 6 feet, complete with hardware, can be furnished at $13.00 extra.

| Catalog Number | Foundation Dimensions Ft. | In. | Ft. | In. | Price | Shipping Weight, Pounds |
|---|---|---|---|---|---|---|
| 55P70 | 9 | 4¾ | x 15 | 1⅞ | $129.00 | 1,215 |
| 55P71 | 9 | 4¾ | x 17 | 7⅞ | 137.00 | 1,325 |
| 55P72 | 9 | 4¾ | x 20 | 1⅞ | 146.00 | 1,440 |

**Double Doors.** Size of opening, 8 ft. by 7 ft. 6 inches. Hinged double doors made of 26-gauge galvanized sheet steel, glazed at top with wire reinforced glass.

**Hinged Double Doors.** Opening, 8 feet by 7 feet 6 inches. 26-gauge galvanized sheet steel, stoutly braced.

**Windows.** One window furnished; fireproof window frame and sash made of 26-gauge galvanized steel, glazed with wire reinforced glass.

## The MERRIMAC — $129.00 And Up

### 55P80 Sectional Fireproof Steel Garage

"The Merrimac." This popular design gable roof steel garage for localities where fire restrictions will not permit a wooden garage. It is sold on the easy payment plan, making it cheaper to buy a garage than to rent it. See our order blank on page 129. Extra windows will be furnished for $6.50 each, including steel frame hinged at top and sash glazed with reinforced wire glass, 16x20 inches. Side door, 2 feet by 6 feet, will be furnished complete with frame and hardware for $13.00 extra.

| Catalog Number | Foundation Dimensions Ft. | In. | Ft. | In. | Price | Shipping Weight, Pounds |
|---|---|---|---|---|---|---|
| 55P80 | 10 | 2¾ | x 12 | 7⅞ | $129.00 | 1,350 |
| 55P81 | 10 | 2¾ | x 15 | 1⅞ | 136.00 | 1,500 |
| 55P82 | 10 | 2¾ | x 17 | 7⅞ | 145.00 | 1,650 |
| 55P83 | 10 | 2¾ | x 20 | 1⅞ | 155.00 | 1,800 |
| 55P84 | 11 | 11 | x 15 | 1⅞ | 146.00 | 1,675 |
| 55P85 | 11 | 11 | x 17 | 7⅞ | 158.00 | 1,875 |
| 55P86 | 11 | 11 | x 20 | 1⅞ | 169.00 | 2,075 |

## The IRONSIDES — $274.00 And Up

### 55P95 Sectional Steel Double Garage

This two-car steel garage is built over a framework of angle irons, making it very strong. It is furnished with two sets of double doors, four windows and service door. Extra windows furnished at $6.50 each. Extra steel doors 2 feet by 6 feet complete with hardware can be furnished for $13.00 each.

**Hinged Double Doors.** Two sets, opening 8 feet by 7 feet 6 inches, 26-gauge galvanized sheet steel.

**Framework.** Sills, wall plates, stringers and uprights are 1x1x⅛-inch angle irons; roof trusses and door frames, 1½x1½x⅛-inch angle irons; the diagonal braces are 1x⅛-inch iron. The roof purlins are 1½x1½x⅛-inch iron angles.

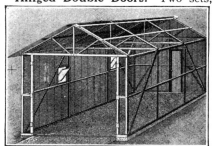

| Catalog Number | Foundation Dimensions Ft. | In. | Ft. | In. | Price | Shipping Weight, Pounds |
|---|---|---|---|---|---|---|
| 55P95 | 15 | 10 | x 17 | 6 | $274.00 | 2,650 |
| 55P97 | 18 | 4 | x 17 | 6 | 310.00 | 2,900 |
| 55P98 | 20 | 0 | x 20 | 0 | 348.00 | 3,450 |
| 55P99 | 23 | 4 | x 20 | 0 | 386.00 | 3,650 |

*Steel Garages Shipped From Factory in OHIO*

# *Simplex Sectional* SUMMER COTTAGES

## SPECIAL OPTIONS

Wallboard for lining, $122.00 extra. Shipping weight, 2,400 pounds.

Screen sections and screen door for 6-foot porch, $47.00 extra.

8-foot porch with screen sections and screen door, $80.00 extra.

Screen doors and half window screens, $20.50 extra.

Batten shutters for windows, $30.00 extra.

### Deductions
If foundation posts are not wanted, deduct $10.00.

## The SILVERHORN

**55P186 Four-Room Sectional House With Two Porches**   **$680.00**
Shipping weight, 12,500 lbs.

Size, 18 feet 3 inches by 24 feet 3 inches

With the gables finished in sea green slate surfaced siding giving a pleasing stucco effect and the attractive paneling of the walls, this four-room summer cottage is distinctive. The arrangement of the rooms is very convenient.

The porch extends across the entire front, 6 feet deep. It can be furnished 8 feet deep at a small extra charge as shown. The bedroom is provided with clothes closet and kitchen with pantry. The kitchen door opens onto a porch 4 feet by 6 feet.

## The WASHINGTON

**55P136 Four-Room Sectional Cottage With Screened Porch**   **$783.00**
Shipping weight, 14,000 lbs.

Size, 32 feet 3 inches by 20 feet 3 inches

The floor plan of this cottage is developed by the U. S. Department of Agriculture to give the greatest economy of floor space. It is really a five-room cottage as the screened porch can be used as a sleeping porch.

Cross ventilation is given to each room, making it very cool. All rooms are light and airy.

Note that the living room is 20 feet long extending entirely through the center of the house.

Price includes screened porch exactly as shown.

## SPECIAL OPTIONS

Wallboard for lining, $138.50 extra. Shipping weight, 3,350 pounds.

Screen doors and half window screens, $16.50 extra.

Batten shutters for windows, $35.50 extra.

### Deductions
If foundation posts are not wanted, deduct $8.00.

## SPECIAL OPTIONS

Wallboard for lining, $194.00 extra.   Shipping weight, 3,950 pounds.

Colonial shutters on all windows, $72.00 extra.

Full length screens for all windows and doors, $48.00 extra.

### Deductions
If foundation posts are not wanted, deduct $13.00.

## The ADIRONDACKS

**55P192 Four-Room Sectional Cottage With Porch and Bathroom**   **$1,146.00**
Shipping weight, 23,000 lbs.

Size, 36 feet 3 inches by 27 feet 4¾ inches

This is our most popular summer cottage, distinctively bungalow in type, with colonial entrance and large French windows. It is furnished with artistic pergolas and colonial columns together with trellises. All rooms are large, well cross ventilated and light.

The 9 feet by 15 feet bedroom may be converted into a garage by placing double doors in front or rear without extra charge. If wanted this way, please advise when ordering.

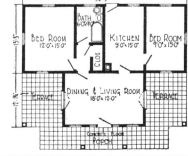

Bathroom fixtures not included in price shown above.

# Simplex Sectional SUMMER COTTAGES

Furnished in two sizes, shown below. Broad sweeping lines, full width porch across front.

Neat, simple and arranged to give the greatest floor area for the least money. All rooms have cross ventilation, insuring plenty of light and fresh air.

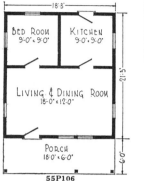

55P106

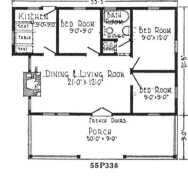

55P338

The Sunburst, 55P106

### SPECIAL OPTIONS

**55P106**
Wallboard for lining, $100.00 extra.
Screen sections for porch with door, $39.50 extra.
Screen doors and half window screens, $17.00 extra.
Batten window shutters, $22.00 extra.

**55P338**
Wallboard for lining, $181.00 extra.
Screen sections for porch with door, $63.00 extra.
Screen doors with half window screens, $27.00 extra.
Batten window shutters, $29.00 extra.

### DEDUCTIONS
If foundation posts are not wanted, deduct $8.00.

If foundation posts are not wanted, deduct $11.50.

This cozy, substantial cottage with low overhanging roof and square

porch columns, is furnished in either of the two plans shown below. Notice how conveniently arranged both plans are. Plenty of fresh air is assured by the many doors and windows.

### SPECIAL OPTIONS

**55P188**
Wallboard for lining $137.50 extra.
Screens for 6-foot porch with doors, $37.50 extra.
Screen doors and half window screens, $22.00 extra.
Batten window shutters, $38.00 extra.

**55P337**
Wallboard for lining, $209.00 extra.
Screen sections for 8-foot porch with doors, $57.00 extra.
Screen doors and half window screens, $26.50 extra.
Batten window shutters, $35.00 extra.

### DEDUCTIONS
If foundation posts are not wanted, deduct $10.00.

If foundation posts are not wanted, deduct $12.00.

55P188

55P337

This cottage is attractively paneled, as shown, with sea green crushed slate surfaced siding on the lower part of the walls and gables, giving a very pretty stucco effect. The rear addition slopes from 8 feet high to 7 feet 6 inches. This cottage is furnished in two different sizes as shown on the plans below.

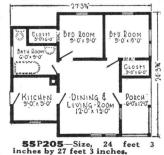

**55P205**—Size, 24 feet 3 inches by 27 feet 3 inches.

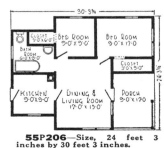

**55P206**—Size, 24 feet 3 inches by 30 feet 3 inches.

### SPECIAL OPTIONS

**55P205**
Wallboard for lining, $152.00 extra.
Screen sections and screen door for porch, $29.50 extra.
Screen doors and half window screens, $19.00 extra.
Batten window shutters, $25.50 extra.

If foundation posts are not wanted, deduct $9.00.

**55P206**
Wallboard for lining $156.00 extra.
Screen sections and screen doors for porch, $35.50 extra.
Screen doors and half window screens, $19.00 extra.
Batten window shutters, $25.50 extra.

### DEDUCTIONS
If foundation posts are not wanted, deduct $9.00.

**For General Specifications See Page 121**

*SEARS, ROEBUCK AND CO.*

# Use This Order Blank for Simplex Sectional Buildings and Garages
## Simplex Sectional Summer Cottages Are Sold for Cash Only

Date_____192___

If you want this order shipped to a town different from your postoffice, give directions here.

Name_____
(SIGN FULL NAME)

Postoffice_____

Rural Route_____ Box No._____ State_____

Street and No._____

**State Total Amount of Money Sent With This Order**

| | DOLLARS | CENTS |
|---|---|---|
| | | |

## Sectional Buildings Only

| Catalog Number of Building Wanted | Quantity Desired | DESCRIPTION | Price of Each |
|---|---|---|---|
| | | | |
| | | | |
| | | | |

## Easy Payment Plan for Garages Only

Send cash payment of $10.00 with order for **One-Car Garage** or $20.00 for **Two-Car Garage.** Balance **equal monthly payments not exceeding eighteen months,** minimum monthly payments, $10.00 on One-Car Garage and $15.00 on Two-Car Garage.

### (Example Showing How Our Garage Easy Payment Plan Works Out)

Price of Garage...................................$204.00  
Payment with order............................ 10.00  } Pay balance at $11.00 per month and interest at 6 per cent.  
Balance.........................................$194.00

Do you own the property on which the garage will be erected?_____How much paid $_____Unpaid $_____

Give description of property **as it reads in your deed**_____

Name of party **holding legal title** (The Title Holder) _____

Cost of property, including improvements, $_____Is property mortgaged?_____State amount, $_____

Describe buildings and other improvements on property_____

State your age_____Are you married?_____Your occupation or business?_____

My family consists of_____

Are you steadily employed?_____How long in present position?_____Weekly earnings, $_____

Give name and address of employer_____

Do you own any other property?_____If so, give value, $_____

Do you have any other income aside from your salary?_____

Mention how long you have lived in town where now located_____ If less than five years give former address_____

### References
(Please Give Names of TWO References)

| NAME | ADDRESS | BUSINESS CR OCCUPATION |
|---|---|---|
| | | |
| | | |

## Agreement
(Answer above questions and fill in and sign this agreement if you wish to order on monthly payments.)

SEARS, ROEBUCK AND CO.

Date_____192___

You may ship me Garage described above. Enclosed is advance payment of $_____ I will pay $_____ monthly with interest at 6 per cent per annum, beginning one month from date shipped, until I have paid in full. Title to and right of possession of the building shall remain in you until I have paid in full.

*Sign Your Name Here* ☞ _____

(If you do not hold legal title to property, the party who does should sign this agreement with you.)

**RENTERS:**—If you are renting and want to put a garage on the property, show our proposition to your landlord and have him sign the order along with you, or if you want the landlord to put up the garage, show him where he can increase the renting power of his property. You will then have the convenience of having the garage right where you live.

# Sears, Roebuck and Co.—*The World's Largest Store*
## CHICAGO    PHILADELPHIA    KANSAS CITY

# Plumbing, Heating, Electric

## FOR "HONOR BILT" MODERN HOMES AND STANDARD BUILT HOUSES
### PRICES ARE SHOWN HERE WITH PAGE REFERENCES TO OTHER PAGES WHERE THE OUTFITS ARE ILLUSTRATED

| MODERN HOME | BATHROOM PLUMBING Illustrated on Page 133 | | HOT WATER HEATING PLANT Illustrated on Page 134 Price, Complete With Boiler, Radiators, Valves, All Necessary Pipe, Pipe Fittings and Pipe Covering | STEAM HEATING PLANT Illustrated on Page 134 Price, Complete With Boiler, Radiators, Valves, All Necessary Pipe, Pipe Fittings and Pipe Covering | WARM AIR HEATING PLANT Illus. on Page 135 Price, Complete With Furnace, Registers, All Necessary Pipe, Pipe Fittings, Galvanized Cold Air Duct and Asbestos Covering | PIPELESS FURNACE Illustrated on Page 135 Price | ELECTRIC WIRING AND FIXTURES KNOB AND TUBE WIRING SYSTEM | | WINDOW SHADES Oil Opaque, See No. 24-1130½ in Our General Catalog Price |
|---|---|---|---|---|---|---|---|---|---|
| | Specification | Price | | | | | Price | Page | Price |
| Albany (13199) | No. 1 | $222.30 | $456.43 | $349.45 | $254.00 | $ 99.80 | $ 90.60 | 138 | $18.00 |
| Albion (3227) | No. 2 | 259.15 | 425.17 | 321.26 | 222.10 | 100.58 | 91.14 | 138 | 19.00 |
| Alhambra (17090) | No. 2 | 275.30 | 608.16 | 422.73 | .......... | .......... | 120.14 | 138 | 38.00 |
| Americus (13063) | No. 2 | 270.50 | 351.29 | 297.38 | 179.35 | 96.80 | 101.38 | 136 | 18.00 |
| Amhurst (3244) | No. 2 | 264.00 | 485.00 | 366.50 | 324.25 | 114.60 | 141.38 | 137 | 19.00 |
| Amsterdam (13196) | No. 2 | 322.30 | 702.32 | 515.17 | .......... | .......... | 139.86 | 137 | 29.00 |
| Ardara (13039) | No. 2 | 249.35 | 432.90 | 340.21 | 178.41 | 93.98 | 85.53 | 138 | 21.00 |
| Ardara With Garage (13039) | No. 2 | 249.35 | 487.05 | 376.15 | 202.91 | .......... | 105.68 | 138 | 21.00 |
| Argyle (17018) | No. 2 | 255.40 | 338.78 | 260.42 | 170.60 | 93.02 | 72.14 | 138 | 17.00 |
| Avalon (13048) | No. 2 | 258.70 | 387.61 | 297.34 | 178.29 | .......... | 99.52 | 136 | 22.00 |
| Barrington (3241) | No. 1 | 180.00 | 460.00 | 360.00 | 257.60 | 100.08 | 142.65 | 137 | 22.00 |
| Bedford (3249) | No. 1 | 218.60 | 522.60 | 393.40 | 272.25 | 113.52 | 116.16 | 138 | 21.00 |
| Betsy Ross (3089) | No. 1 | 192.00 | 276.79 | 198.45 | 129.65 | 76.83 | 59.69 | 138 | 13.00 |
| Chesterfield (3235) | No. 2 | 231.50 | 607.05 | 457.30 | 194.45 | 84.35 | 178.05 | 137 | 25.00 |
| Cleveland (3233) | No. 1 | 350.00 | 389.60 | 295.43 | 280.00 | .......... | 102.12 | 138 | 22.00 |
| Clyde (9030) | No. 1 | 209.35 | 302.49 | 229.14 | 152.65 | 92.06 | 68.21 | 138 | 13.00 |
| Columbine (8013) | No. 2 | 249.65 | 350.22 | 266.89 | 186.00 | .......... | 76.54 | 138 | 16.00 |
| Columbine With Attic (8013) | No. 2 | 249.65 | 431.91 | 310.07 | 211.00 | .......... | 96.10 | 138 | 20.00 |
| Conway (13052) | No. 1 | 204.65 | 276.43 | 223.63 | 168.00 | 92.06 | 70.01 | 138 | 10.00 |
| Conway With Attic (13052) | No. 1 | 204.65 | 316.87 | 248.41 | 214.50 | 98.36 | 93.77 | 138 | 12.00 |
| Cornell (3226) | No. 1 | 211.40 | 387.06 | 292.83 | 172.39 | 81.57 | 58.14 | 139 | 18.00 |
| Crescent (13084) | No. 2 | 259.80 | 391.90 | 276.96 | 167.91 | 106.04 | 69.11 | 138 | 19.00 |
| Crescent With Attic (13084) | No. 2 | 259.80 | 477.45 | 331.35 | 215.80 | 109.34 | 93.09 | 138 | 22.00 |
| Crescent (13086) | No. 2 | 266.80 | 352.19 | 268.71 | 182.08 | 107.00 | 66.72 | 138 | 15.00 |
| Crescent With Attic (13086) | No. 2 | 266.80 | 408.93 | 302.53 | 198.22 | 110.30 | 87.15 | 138 | 20.00 |
| Del Rey (13065) | No. 2 | 248.50 | 369.19 | 304.48 | .......... | .......... | 108.81 | 136 | 19.00 |
| Dundee (3209) | No. 1 | 201.00 | 233.19 | 177.39 | 122.85 | 78.27 | 57.49 | 138 | 10.00 |
| Elsmore (13192) | No. 2 | 264.15 | 309.32 | 257.50 | .......... | .......... | 81.87 | 136 | 16.00 |
| Estes (6014) | .......... | 45.20 | .......... | .......... | 117.90 | 66.70 | 28.81 | 139 | 9.00 |
| Fairy (3217) | No. 1 | 201.50 | 224.46 | 174.39 | 127.70 | 66.70 | 35.08 | 139 | 9.00 |
| Farnum (6017) | .......... | 151.75 | .......... | .......... | 122.60 | 67.58 | 34.45 | 139 | 10.00 |
| Fosgate (6016) | .......... | 149.00 | .......... | .......... | 111.40 | 66.70 | 25.33 | 139 | 8.00 |
| Fullerton (3205) | No. 1 | 210.65 | 401.83 | 307.71 | 194.75 | 96.32 | 99.20 | 138 | 19.00 |
| Garfield (3232) | No. 1 | 391.60 | 474.97 | 380.21 | 255.70 | .......... | 142.87 | 138 | 23.00 |
| Glen Falls (3245) | No. 2 | 405.25 | 655.00 | 551.00 | 396.10 | .......... | 182.87 | 137 | 26.00 |
| Gladstone (3222) | No. 2 | 271.60 | 402.76 | 295.76 | 172.28 | 82.05 | 84.36 | 138 | 18.00 |
| Grant (6018) | .......... | 154.00 | .......... | .......... | 148.00 | 78.75 | 41.35 | 139 | 11.00 |
| Hamilton (3200) | No. 1 | 187.10 | 369.33 | 278.13 | 182.70 | 93.50 | 74.28 | 138 | 17.00 |
| Hampton (3208) | No. 1 | 203.15 | 302.18 | 239.07 | 146.55 | 68.46 | 50.37 | 139 | 13.00 |
| Hathaway (3195) | No. 1 | 214.80 | 381.19 | 293.51 | 160.10 | 96.32 | 81.75 | 138 | 26.00 |
| Homewood (3238) | No. 1 | 188.10 | 506.00 | 373.00 | 326.55 | 113.52 | 135.64 | 137 | 19.00 |
| Honor (13071) | No. 2 | 351.00 | 641.18 | 552.78 | 263.10 | 108.86 | 123.62 | 136 | 34.00 |
| Hudson (6013) (6013A) | .......... | 154.85 | .......... | .......... | 124.75 | 66.70 | 25.72 | 139 | 8.00 |
| Josephine (7044) | No. 1 | 199.00 | 239.01 | 183.15 | 127.10 | 77.31 | 41.16 | 139 | 10.00 |
| Kilbourne (17013) | No. 2 | 258.25 | 385.93 | 304.51 | 189.95 | 106.04 | 82.32 | 136 | 17.00 |
| Kilbourne With Attic (17013) | No. 2 | 258.25 | 487.64 | 374.43 | 239.90 | 109.34 | 119.90 | 136 | 24.00 |
| Kimball (6015) | .......... | 47.30 | .......... | .......... | 102.00 | 66.70 | 21.85 | 139 | 9.00 |
| Kismet (17002) | No. 1 | 208.35 | 209.83 | 172.61 | 127.10 | 78.27 | 45.33 | 139 | 9.00 |
| La Salle (3243) | No. 1 | 406.50 | 531.70 | 415.00 | 412.35 | .......... | 121.69 | 138 | 21.00 |

**NOTE—Prices for plumbing as given above include fixtures specified on page mentioned, together with all necessary pipe and fittings required to install same to the point where the soil pipe stack comes straight down and intersects with the basement floor. It is understood that all galvanized iron supply pipe and cast iron soil pipe which may be required to make connection from this point on are to be charged for extra. A special estimate will be furnished for this extra material on request.**

In the case of those few houses in the above list where the price of the pipeless furnace or warm air heating system is not extended, we recommend hot water or steam system. If there are any special local or city requirements in regard to the manner of wiring or plumbing or the use of certain material and its installation it will be necessary for you to give us this information.

P598

SEARS, ROEBUCK AND CO., PHILADELPHIA

# Lighting and Window Shades

## FOR "HONOR BILT" MODERN HOMES AND STANDARD BUILT HOUSES
### PRICES ARE SHOWN HERE WITH PAGE REFERENCES TO OTHER PAGES WHERE THE OUTFITS ARE ILLUSTRATED

| MODERN HOME | BATHROOM PLUMBING Illustrated on Page 133 | | HOT WATER HEATING PLANT Illustrated on Page 134 Price, Complete With Boiler, Radiators, Valves, All Necessary Pipe, Pipe Fittings and Pipe Covering | STEAM HEATING PLANT Illustrated on Page 134 Price, Complete With Boiler, Radiators, Valves, All Necessary Pipe, Pipe Fittings and Pipe Covering | WARM AIR HEATING PLANT Illus. on Page 135 Price, Complete With Furnace, Registers, All Necessary Pipe, Pipe Fittings, Galvanized Cold Air Duct and Asbestos Covering | PIPELESS FURNACE Illustrated on Page 135 | ELECTRIC WIRING AND FIXTURES KNOB AND TUBE WIRING SYSTEM | | WINDOW SHADES Oil Opaque, See No. 24-1130½ in Our General Catalog |
|---|---|---|---|---|---|---|---|---|---|
| | Specification | Price | | | | Price | Price | Page | Price |
| Lexington (13045)........ | No. 2 | $351.80 | $708.63 | $574.01 | ............ | ............ | $193.87 | 137 | $48.00 |
| Manchester (3250)....... | No. 1 | 362.10 | 534.30 | 428.30 | $369.25 | ............ | 125.07 | 138 | 22.00 |
| Martha Washington (13080)........ | No. 2 | 231.75 | 562.51 | 426.46 | ............ | ............ | 127.75 | 137 | 44.00 |
| Montrose (3239)........ | No. 2 | 277.00 | 685.50 | 455.38 | 331.70 | $114.64 | 144.96 | 137 | 29.00 |
| Norwood (2095)........ | No. 1 | 213.00 | 310.24 | 231.91 | 144.35 | 82.05 | 51.33 | 139 | 15.00 |
| Oakdale (3206).......... | No. 1 | 198.30 | 325.85 | 256.95 | 152.45 | 77.35 | 65.83 | 138 | 14.00 |
| Oak Park (3237A)....... | No. 2 | 261.60 | 545.15 | 433.00 | 345.85 | 113.52 | 150.17 | 137 | 23.00 |
| Oak Park (3237B)....... | No. 2 | 262.70 | 570.60 | 429.15 | 314.20 | ............ | 181.18 | 137 | 23.00 |
| Olivia (7028)........... | No. 1 | 208.25 | 225.95 | 181.34 | 130.00 | 66.70 | 41.81 | 139 | 9.00 |
| Osborn (12050)......... | No. 2 | 251.35 | 419.14 | 331.37 | 181.45 | 93.98 | 127.94 | 136 | 18.00 |
| Prescott (3240)......... | No. 1 | 172.65 | 304.00 | 242.20 | 200.65 | 94.21 | 79.04 | 138 | 11.00 |
| Priscilla (3229)........ | No. 2 | 236.65 | 419.90 | 340.26 | 192.45 | 112.82 | 132.04 | 137 | 27.00 |
| Puritan With Sun Room (3190A)..... | No. 1 | 209.00 | 500.48 | 361.32 | 253.70 | ............ | 121.36 | 137 | 30.00 |
| Puritan (3190B)......... | No. 1 | 209.00 | 424.92 | 329.78 | 248.30 | ............ | 127.54 | 137 | 19.00 |
| Ramsay (6012)......... | | 146.15 | ............ | ............ | 110.65 | 66.70 | 26.09 | 139 | 8.00 |
| Rembrandt With Sun Room (3215A)..... | No. 2 | 269.30 | 531.63 | 397.56 | ............ | ............ | 133.59 | 137 | 30.00 |
| Rembrandt (3215B)...... | No. 2 | 269.30 | 520.74 | 353.51 | ............ | ............ | 126.53 | 137 | 19.00 |
| Rodessa (3203)......... | | 48.00 | 217.30 | 179.76 | 119.30 | 68.02 | 49.11 | 138 | 9.00 |
| Rodessa (7041)......... | No. 1 | 206.60 | 252.39 | 182.79 | 122.95 | 68.02 | 52.46 | 138 | 9.00 |
| Rockford (3251)........ | No. 1 | 212.50 | 450.10 | 364.00 | 279.00 | 113.06 | 100.96 | 138 | 21.00 |
| Salem (3211)........... | No. 1 | 216.20 | 508.34 | 373.08 | 207.65 | 95.21 | 132.95 | 137 | 18.00 |
| Selby (6011)........... | | 156.00 | ............ | ............ | 110.65 | 66.70 | 26.04 | 139 | 9.00 |
| Sheridan (3224)........ | No. 1 | 204.00 | 303.96 | 246.81 | 165.70 | ............ | 72.16 | 138 | 13.00 |
| Sheridan With Attic (3224)....... | No. 1 | 204.00 | 383.37 | 287.31 | 208.10 | ............ | 102.65 | 138 | 17.00 |
| Solace (3218).......... | No. 1 | 204.00 | 363.58 | 280.08 | 154.43 | 82.80 | 53.53 | 139 | 15.00 |
| Somers (7008)......... | No. 1 | 193.00 | 300.05 | 248.61 | 161.23 | 68.46 | 78.79 | 138 | 13.00 |
| Starlight (3202)........ | | 49.40 | 238.70 | 193.34 | 125.00 | 78.25 | 60.72 | 138 | 11.00 |
| Starlight (7009)........ | No. 1 | 198.60 | 241.85 | 195.06 | 151.90 | 78.27 | 68.34 | 138 | 13.00 |
| Sunbeam (3194)........ | No. 2 | 259.55 | 454.36 | 336.97 | 158.50 | 97.13 | 108.61 | 137 | 19.00 |
| Sunlight (3221)........ | No. 1 | 200.00 | 260.63 | 198.76 | 133.50 | 78.17 | 71.41 | 138 | 12.00 |
| Tarryton (3247)........ | No. 2 | 360.00 | 571.00 | 449.00 | 307.40 | 112.15 | 175.70 | 137 | 25.00 |
| Vallonia (13049)........ | No. 2 | 251.80 | 295.13 | 241.90 | 148.25 | 93.50 | 72.31 | 138 | 13.00 |
| Vallonia With Attic (13049)...... | No. 2 | 251.80 | 365.96 | 283.13 | 181.85 | 96.65 | 97.46 | 138 | 18.00 |
| Van Dorn (3234)....... | No. 1 | 209.40 | 387.02 | 291.62 | 219.75 | 85.27 | 119.81 | 137 | 16.00 |
| Van Page (3242)....... | No. 1 | 224.65 | 490.35 | 376.80 | 287.90 | 100.94 | 147.72 | 137 | 22.00 |
| Verona (13201)......... | No. 2 | 351.90 | 674.25 | 489.49 | ............ | ............ | 170.97 | 136 | 29.00 |
| Walton (13050)........ | No. 2 | 259.00 | 386.74 | 272.11 | 203.51 | 93.50 | 107.08 | 138 | 16.00 |
| Wayne (13210).......... | No. 2 | 260.40 | 351.54 | 262.98 | 164.44 | 98.36 | 81.73 | 138 | 20.00 |
| Wellington (3223)....... | No. 2 | 251.35 | 277.85 | 232.84 | 174.58 | 78.75 | 71.69 | 138 | 13.00 |
| Westly (13085).......... | No. 2 | 250.00 | 384.43 | 320.33 | 245.44 | 109.67 | 93.77 | 138 | 24.00 |
| Whitehall (3035)........ | No. 1 | 211.20 | 370.26 | 256.45 | 167.37 | 80.94 | 83.26 | 138 | 19.00 |
| Windemere (1208)....... | No. 1 | 389.50 | 600.00 | 468.23 | 308.20 | ............ | 136.21 | 138 | 30.00 |
| Windsor (3193)......... | No. 1 | 218.00 | 292.52 | 226.50 | 147.95 | 80.94 | 75.67 | 138 | 13.00 |
| Winona (2010).......... | No. 1 | 209.80 | 267.03 | 203.85 | 143.20 | 93.50 | 71.37 | 138 | 12.00 |
| Winona (2010B)......... | No. 1 | 204.45 | 316.43 | 252.18 | 152.69 | 93.50 | 73.48 | 138 | 14.00 |
| Woodland (3025)....... | No. 2 | 271.00 | 423.03 | 316.77 | 257.25 | 96.65 | 121.50 | 138 | 18.00 |
| Sun Room (3213)........ | | ............ | ............ | ............ | ............ | ............ | 3.25 | 137 | 12.00 |
| Sun Room (3214)........ | | ............ | ............ | ............ | ............ | ............ | 3.25 | 137 | 12.00 |

NOTE—Prices for plumbing as given above include fixtures specified on page mentioned, together with all necessary pipe and fittings required to install same to the point where the soil pipe stack comes straight down and intersects with the basement floor. It is understood that all galvanized iron supply pipe and cast iron soil pipe which may be required to make connection from this point on are to be charged for extra. A special estimate will be furnished for this extra material on request.

In the case of those few houses in the above list where the price of the pipeless furnace or warm air heating system is not extended, we recommend hot water or steam system. If there are any special local or city requirements in regard to the manner of wiring or plumbing or the use of certain material and its installation it will be necessary for you to give us this information.

# High Grade Building Hardware
## for Modern Homes

## Chicago Design

Front and rear door lock sets are made of genuine brass. We usually furnish this design in old copper finish, but same can be furnished in lemon brass finish if desired.

Inside door lock sets are made of electro plated steel and finished in either old copper or lemon brass.

Bathroom door lock set is made of genuine brass with either old copper or lemon brass finish on the outside, and nickel plated finish inside to match the balance of the bathroom hardware.

Other exposed articles have either old copper or lemon brass finish, thus securing a uniform treatment throughout the house.

Push plates for swinging doors are made of heavy beveled edge plate glass. Nails of all sizes and kinds required are furnished.

For those who desire solid brass hardware throughout their house, we are in a position to furnish this material and will be glad to give prices upon request.

## Stratford Design

Front and rear door lock sets are made of genuine brass. We usually furnish this design in lemon brass finish, but can also supply old copper finish if desired.

Inside door lock sets are made of electro plated steel and finished in either lemon brass or old copper.

Bathroom lock set is made of genuine brass with either lemon brass or old copper on the outside, and nickel plated finish inside to match the balance of the bathroom hardware.

Other exposed articles have lemon brass or old copper finish, thus giving a uniform treatment throughout the house.

Push plates for swinging doors are made of heavy beveled edge plate glass. Nails of all sizes and kinds required are furnished.

For those who desire solid brass hardware throughout their house, we are in a position to furnish this material and will be glad to give prices upon request.

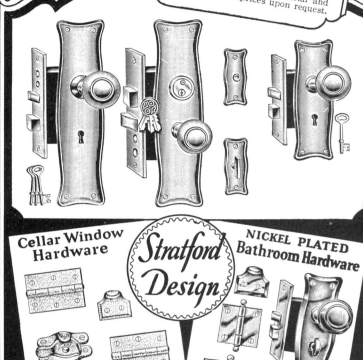

### NICKEL PLATED Bathroom Hardware — Chicago Design — Cellar Window Hardware

### Cellar Window Hardware — Stratford Design — NICKEL PLATED Bathroom Hardware

### Cupboard Hardware

### Casement Hardware

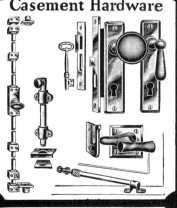

### Casement Hardware

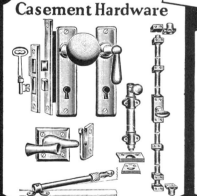

### Cupboard Hardware

### Swinging Door Hardware

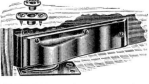

### Swinging Door Hardware

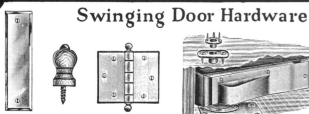
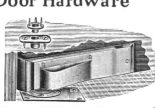

# Plumbing Systems Furnished With Modern Homes

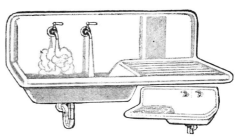

**This is our Delaware One-Piece Cast Iron White Porcelain Enameled Kitchen Sink, complete with faucets and trap to wall. This sink is included in Specification No. 1.**

### Complete Plumbing System Specification No. 1

Includes Fairview bathroom outfit, as shown above at left. Bathtub is painted white on outside, porcelain enameled inside and over rim, a 30-gallon galvanized range boiler, two-compartment laundry tub, size 48x24 inches, as shown in center of page, one-piece sink, with drain board, as shown above, and all necessary cast iron soil pipe, calking lead, oakum, galvanized iron water supply and waste pipe, galvanized fittings, faucets, traps and connections; in fact, everything required to install the entire plumbing system in a first class, sanitary manner from point where soil pipe stack comes straight down and touches basement floor to roof of building. Piping required beyond this point not included in prices given but separate estimate will be furnished on this extra piping to make sewer and water connection to a point any distance away from building on request.

We furnish plans and full instructions for installing. Any handy man can easily do the work. All connections can be made with threaded iron pipe. System is guaranteed to give satisfactory service. Gas piping or pneumatic water supply system not included. Separate estimate on these furnished on request.

NOTE—Ordinances of various cities differ widely, some requiring more and some less pipe and fittings. In the above we have provided a good sanitary plumbing system, but it may not comply with the ordinances of your particular city. If you are required to follow special city ordinances it is better to send us a copy of these ordinances and get our special estimate before ordering.

NOTE—Laundry Tub, Range Boiler, Soil Pipe, Fittings, etc., as illustrated in this center panel, are included in both Specifications No. 1 and No. 2.

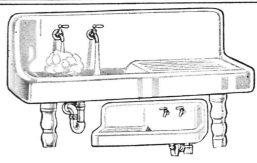

**This is our Hiawatha One-Piece Cast Iron White Porcelain Enameled Apron Sink, fitted complete with painted adjustable cast iron legs, faucets and trap. This sink is included in Specification No. 2.**

### Complete Plumbing System Specification No. 2

Includes De Luxe bathroom outfit, as shown at left. Bathtub is painted white on outside, porcelain enameled inside and over rim. 30-gallon galvanized range boiler, two-compartment laundry tub, size 48x24 inches, as shown in center of page, and cast iron enameled one-piece apron sink, as shown above, with all necessary galvanized iron water supply and waste pipe, soil pipe, calking lead, oakum, galvanized fittings, faucets, traps and connections, etc.; in fact, everything required to install the entire system in a first class, sanitary manner from point where soil pipe stack comes straight down and touches floor line in basement to roof of building. Separate estimate for extra cost of water supply pipe and cast iron soil pipe to make connections to sewer and to source of water supply from point where pipes come down to ground line in basement furnished on request.

Gas piping or pneumatic water supply system not included. Separate estimate on these furnished on request.

System guaranteed to give satisfactory service. Complete plans and instructions for installing furnished free.

NOTE—Ordinances of various cities differ widely, some requiring more and some less pipe and fittings. In the above we have provided a good sanitary plumbing system, but it may not comply with the ordinances of your particular city. If you are required to follow special city ordinances it is better to send us a copy of these ordinances and get our special estimate before ordering.

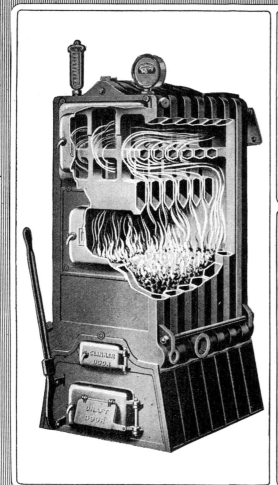

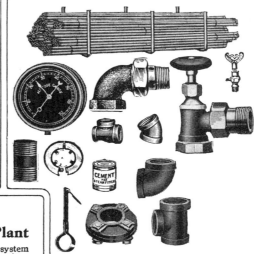

## Hercules
## Hot Water Heating Plant

Complete hot water heating system consists of Hercules boiler and cast iron radiators with all necessary pipe, valves, fittings, etc., required to completely install, galvanized expansion tank, and all necessary covering for boiler and pipes in basement. Includes also complete plans and instructions for installing. Radiators to be of heights and sizes as indicated in our specifications furnished on request.

System guaranteed to keep house comfortably warm and to give satisfactory service in coldest winter weather.

Special estimate will be furnished for heating system without pipe covering on request. This covering may be temporarily omitted and applied later, but it is understood that to comply with our guarantee and to get full measure of efficiency out of your heating system, pipes in basement must be covered.

Water heating coil for boiler to heat water for domestic use included without extra charge.

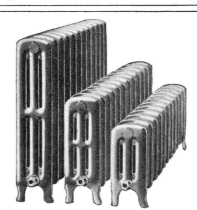

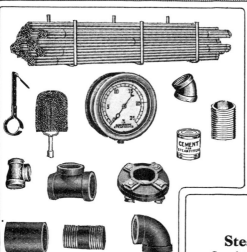

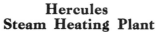

## Hercules
## Steam Heating Plant

Complete steam heating plant consists of Hercules cast iron boiler, cast iron radiators with all necessary pipe, valves, fittings, etc., required to completely install, all necessary steam trimmings for boiler as illustrated, and all necessary covering for boiler and pipes in basement. Radiators to be of heights and sizes as indicated in our specifications furnished on request. Water heating coil for boiler to heat water for domestic use included without extra charge. Complete plans and instructions for installing furnished.

System guaranteed to keep house comfortably warm in coldest winter weather.

Special estimate will be furnished for heating system without pipe covering on request. This covering may be temporarily omitted and applied later, but it is understood that to comply with our guarantee and to get full measure of efficiency out of your heating system, pipes in basement must be covered.

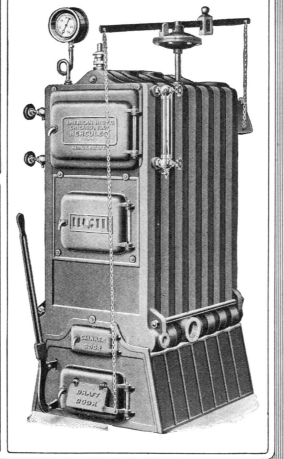

Floor Register.

Baseboard Register.

Side Wall Register for Above Baseboard.

NOTE—We specify wall registers in all cases except where structural design at any particular point in a building makes it absolutely necessary to use a floor register.

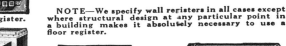

Wood Cold Air Face.

Showing Our Lock Seam Warm Air Pipe.

Box for Side Wall Register.

Baseboard Register Box.

Stack Elbow.

Safety Wall Pipe.

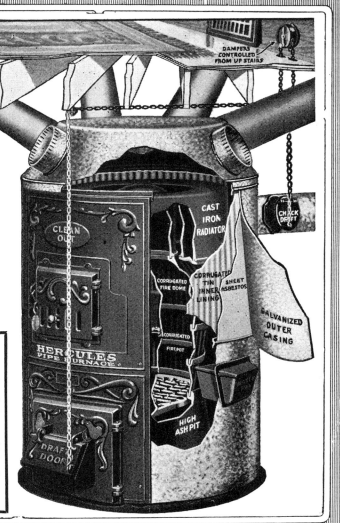

## Hercules Warm Air Heating System

Complete warm air heating plant includes Hercules Warm Air Pipe Furnace with all necessary tin furnace pipe, galvanized smoke pipe, tin register boxes, black japanned steel registers and asbestos paper for covering pipes. We furnish only first quality double construction tin wall boxes, boots, wall stack and fittings on all our Hercules warm air heating systems. Furnace is well made of cast iron with heavily corrugated fire pot and fire dome castings, high ash pit and revolving triangular grates. Galvanized sheet metal cold air connection included. We furnish plans and instructions for installing and all holes are cut in top bonnet of furnace before shipment to connect warm air pipes according to plan.

A hot water coil to be inserted in fire pot of furnace for heating water for domestic use is included without extra charge.

Entire system guaranteed to give satisfactory service.

## The Hercules Pipeless Furnace

Our Hercules Pipeless Furnace embodies all modern improvements in pipeless furnace design. Furnace proper is made of cast iron throughout. The fire pot and fire dome castings are heavily corrugated and have deep cup joints, which prevent gas or smoke escaping into the warm air chamber.

### Revolving Triangular Grates
*Easy to Operate*

Grate bars are of the revolving triangular pattern, controlled by a simple mechanism, which makes the grates easy to revolve, and a woman or a boy can easily operate them.

### Cold Air Is Taken From Upstairs

By carefully studying the illustration of our Hercules Pipeless Furnace it will be observed that the cold air is taken from upstairs. It goes down through the large register around the outside edges and turns up inside of the inner casing, and after passing over the hot furnace is discharged through the center of the large register. The furnace is well made and guaranteed to give satisfactory service.

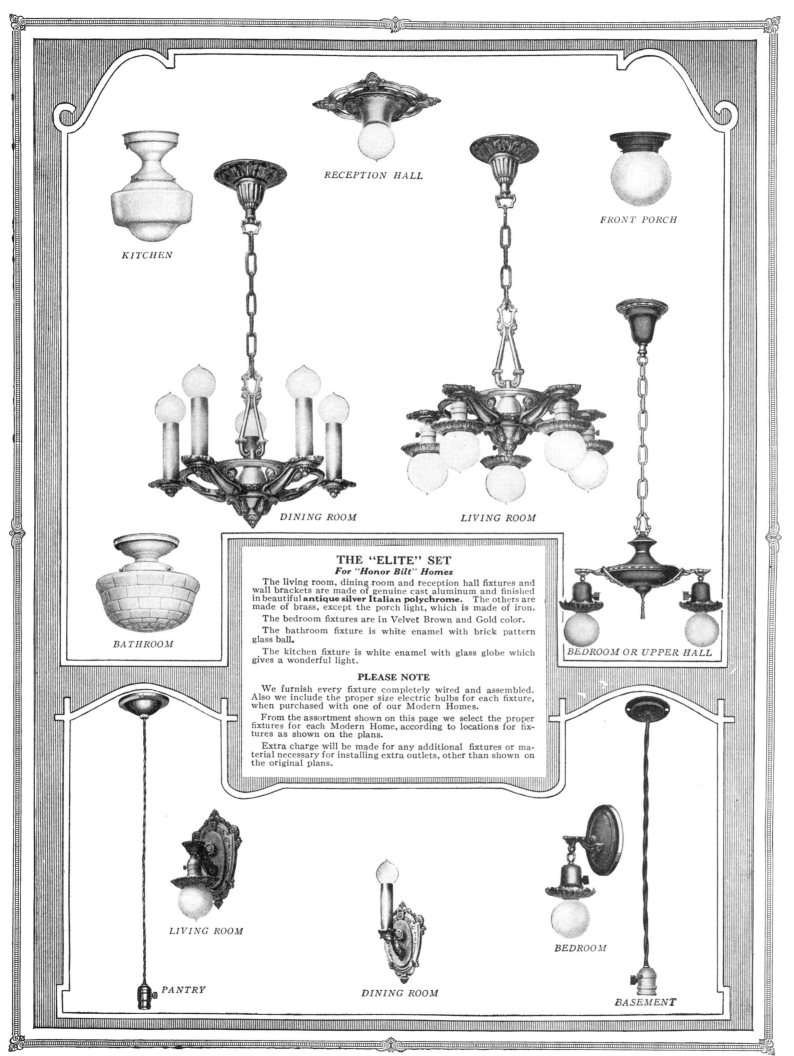

RECEPTION HALL

KITCHEN

FRONT PORCH

DINING ROOM

LIVING ROOM

BATHROOM

### THE "ELITE" SET
*For "Honor Bilt" Homes*

The living room, dining room and reception hall fixtures and wall brackets are made of genuine cast aluminum and finished in beautiful **antique silver Italian polychrome.** The others are made of brass, except the porch light, which is made of iron.

The bedroom fixtures are in Velvet Brown and Gold color.

The bathroom fixture is white enamel with brick pattern glass ball.

The kitchen fixture is white enamel with glass globe which gives a wonderful light.

#### PLEASE NOTE

We furnish every fixture completely wired and assembled. Also we include the proper size electric bulbs for each fixture, when purchased with one of our Modern Homes.

From the assortment shown on this page we select the proper fixtures for each Modern Home, according to locations for fixtures as shown on the plans.

Extra charge will be made for any additional fixtures or material necessary for installing extra outlets, other than shown on the original plans.

BEDROOM OR UPPER HALL

LIVING ROOM

PANTRY

DINING ROOM

BEDROOM

BASEMENT

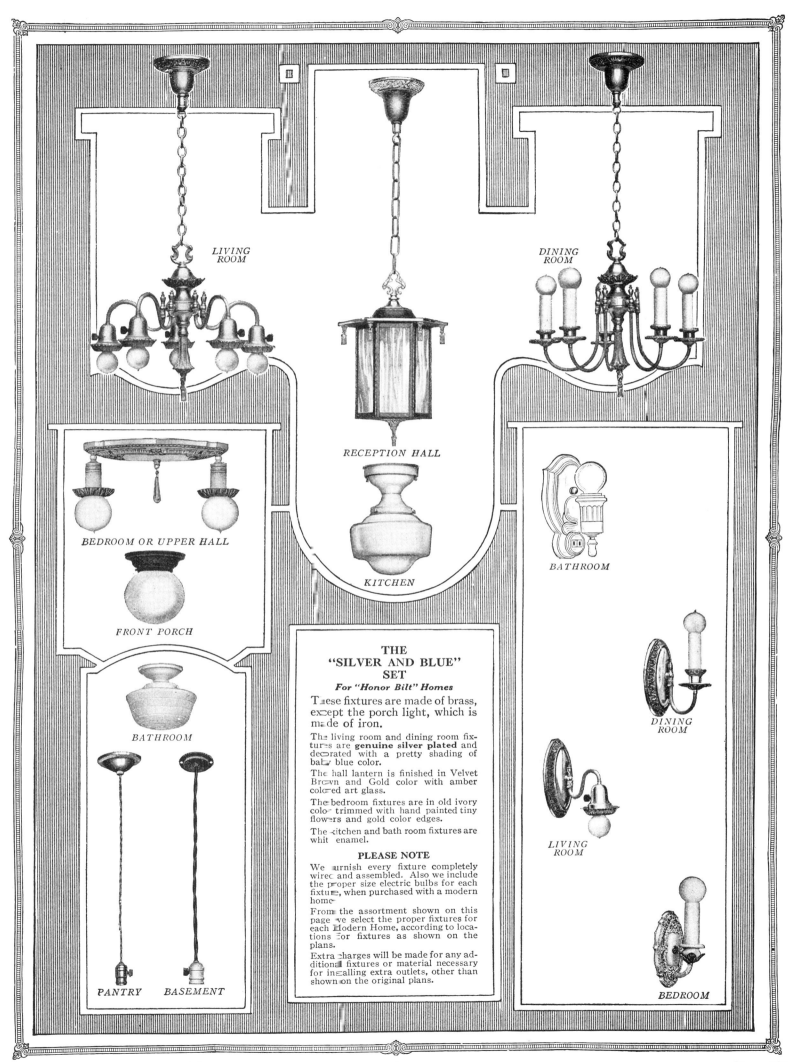

LIVING ROOM

DINING ROOM

RECEPTION HALL

BEDROOM OR UPPER HALL

FRONT PORCH

BATHROOM

KITCHEN

BATHROOM

BATHROOM

DINING ROOM

LIVING ROOM

PANTRY    BASEMENT

BEDROOM

## THE "SILVER AND BLUE" SET

### For "Honor Bilt" Homes

These fixtures are made of brass, except the porch light, which is made of iron.

The living room and dining room fixtures are **genuine silver plated** and decorated with a pretty shading of baby blue color.

The hall lantern is finished in Velvet Brown and Gold color with amber colored art glass.

The bedroom fixtures are in old ivory color trimmed with hand painted tiny flowers and gold color edges.

The kitchen and bath room fixtures are whit enamel.

### PLEASE NOTE

We furnish every fixture completely wired and assembled. Also we include the proper size electric bulbs for each fixture, when purchased with a modern home.

From the assortment shown on this page we select the proper fixtures for each Modern Home, according to locations for fixtures as shown on the plans.

Extra charges will be made for any additional fixtures or material necessary for installing extra outlets, other than shown on the original plans.

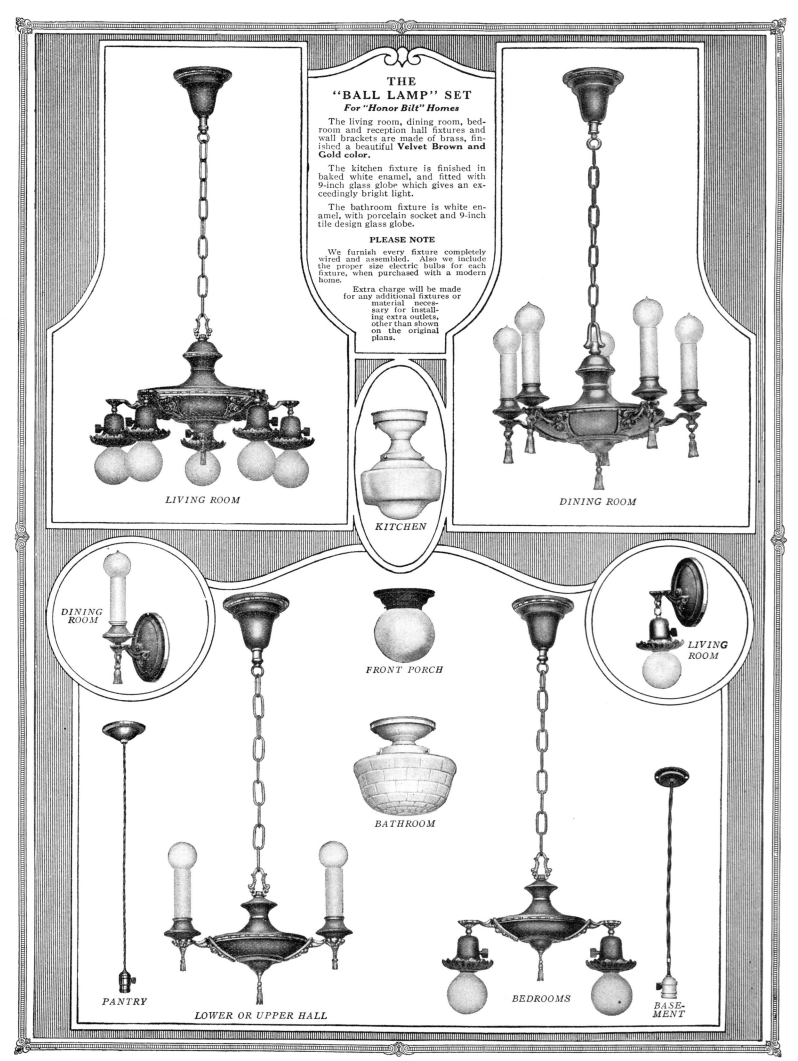

# THE "BALL LAMP" SET
### For "Honor Bilt" Homes

The living room, dining room, bedroom and reception hall fixtures and wall brackets are made of brass, finished a beautiful **Velvet Brown and Gold color.**

The kitchen fixture is finished in baked white enamel, and fitted with 9-inch glass globe which gives an exceedingly bright light.

The bathroom fixture is white enamel, with porcelain socket and 9-inch tile design glass globe.

**PLEASE NOTE**

We furnish every fixture completely wired and assembled. Also we include the proper size electric bulbs for each fixture, when purchased with a modern home.

Extra charge will be made for any additional fixtures or material necessary for installing extra outlets, other than shown on the original plans.

LIVING ROOM

KITCHEN

DINING ROOM

DINING ROOM

FRONT PORCH

LIVING ROOM

BATHROOM

PANTRY

LOWER OR UPPER HALL

BEDROOMS

BASEMENT

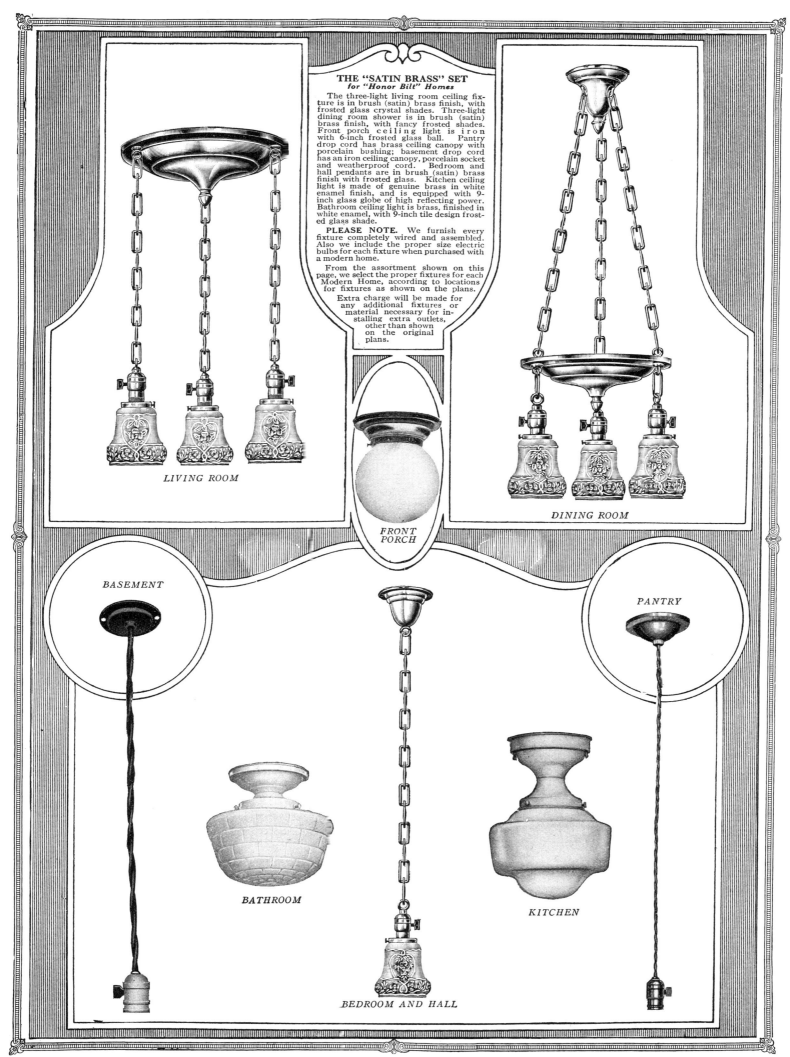

### THE "SATIN BRASS" SET
#### for "Honor Bilt" Homes

The three-light living room ceiling fixture is in brush (satin) brass finish, with frosted glass crystal shades. Three-light dining room shower is in brush (satin) brass finish, with fancy frosted shades. Front porch ceiling light is iron with 6-inch frosted glass ball. Pantry drop cord has brass ceiling canopy with porcelain bushing; basement drop cord has an iron ceiling canopy, porcelain socket and weatherproof cord. Bedroom and hall pendants are in brush (satin) brass finish with frosted glass. Kitchen ceiling light is made of genuine brass in white enamel finish, and is equipped with 9-inch glass globe of high reflecting power. Bathroom ceiling light is brass, finished in white enamel, with 9-inch tile design frosted glass shade.

**PLEASE NOTE.** We furnish every fixture completely wired and assembled. Also we include the proper size electric bulbs for each fixture when purchased with a modern home.

From the assortment shown on this page, we select the proper fixtures for each Modern Home, according to locations for fixtures as shown on the plans.

Extra charge will be made for any additional fixtures or material necessary for installing extra outlets, other than shown on the original plans.

LIVING ROOM

FRONT PORCH

DINING ROOM

BASEMENT

PANTRY

BATHROOM

KITCHEN

BEDROOM AND HALL

# Guaranteed *Slate Surfaced* ASPHALT SHINGLES

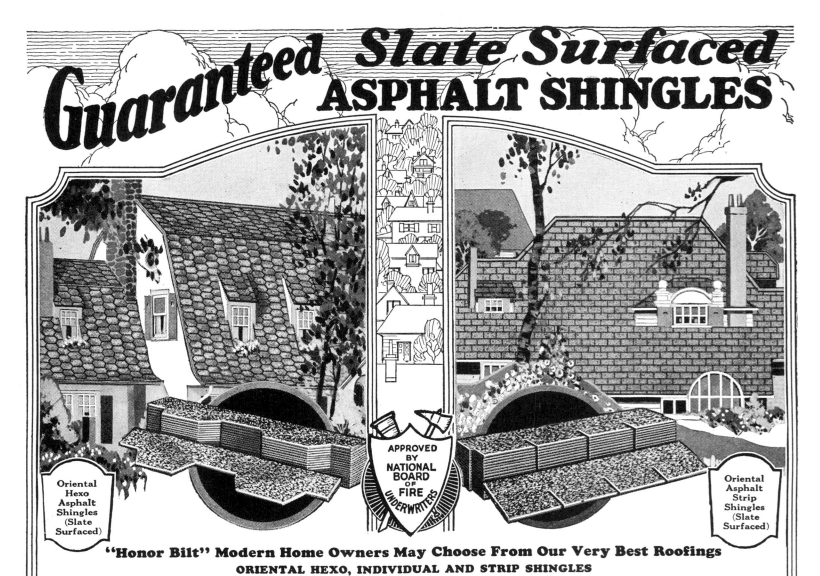

Oriental Hexo Asphalt Shingles (Slate Surfaced)

APPROVED BY NATIONAL BOARD OF FIRE UNDERWRITERS

Oriental Asphalt Strip Shingles (Slate Surfaced)

## "Honor Bilt" Modern Home Owners May Choose From Our Very Best Roofings
### ORIENTAL HEXO, INDIVIDUAL AND STRIP SHINGLES
*Guaranteed 17 Years*
*Approved by National Board of Fire Underwriters*

"Honor Bilt" home owners and contractors warmly praise our three styles of Oriental Slate Surfaced Shingles, illustrated above and below. Of course, a beautiful, shingled roof home always takes on a higher valuation.

Oriental Shingles are the perfected result of twenty-five years' manufacturing experience. Fabricated of highest quality materials, known for long wearing qualities, including heavy roofing felt, imported asphalt and natural color, red or green slate.

You may choose any style of Oriental Shingles for your "Honor Bilt" Home, confident in the knowledge that they represent the utmost in value. They are rigid, lasting, economical and fire resisting; will not decay, rust, warp, curl or blow in the wind. Our shingles defy all weather conditions, such as rain, sleet and snow, the rigorous blasts of winter or the torrid heat of summer. Makes the home warmer in winter.

The Following "Honor Bilt" Homes Are Specified With Oriental Individual Asphalt Shingles (Slate Surfaced).

| No. | Name | No. | Name |
|---|---|---|---|
| 3227 | —Albion | 13045 | —Lexington |
| 17090A | —Alhambra | 13080A | —Martha Washington |
| 13063 | —Americus | | |
| 13039 | —Ardara | 3215A | —Rembrandt |
| 13199 | —Albany | 3215B | —Rembrandt |
| 13048 | —Avalon | 3211 | —Salem |
| 13086A | —Crescent | 3324 | —Sheridan |
| 13084A | —Crescent | 3229 | —Priscilla |
| 13065 | —Del Rey | 3194A | —Sunbeam |
| 3205 | —Fullerton | 13210 | —Wayne |
| 13071 | —Honor | | |

NOTE—Oriental Hexo Shingles or Oriental Strip Shingles can be furnished. Write for information.

## APPROVED BY FIRE UNDERWRITERS ~ GUARANTEED 17 YEARS

### Additional Information

Regarding "Honor Bilt" Homes, such as material, construction, millwork, and manufacturing facilities, including our Free Architectural Service, will be found on other pages of this book.

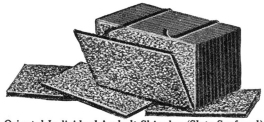

**Oriental Individual Asphalt Shingles (Slate Surfaced)**

The artistic and beautiful appearance of houses roofed with Oriental Slate Surfaced Shingles is increasing their value everywhere. The color never fades and needs no paint or stain. Practically fireproof. Made of the same high grade felt and asphalt as our Oriental Slate Surfaced Roofing, except that our Oriental Shingles contain an especially treated, imported asphalt that stiffens the shingles, so that they will lay perfectly flat.

### Interesting and Valuable

Information about our Easy Payment Plan of Modern Homes will be found on the inside front cover and on page 144. Order Blank on page 143. Information Blank on page 141. Your next step on page 142.

# Information Blank

## USE FOR MODERN HOMES Only

## NO OBLIGATION NO EXPENSE

## USE FOR MODERN HOMES Only

You took the first step toward owning your home when you wrote for this Book of "Honor Bilt" Modern Homes. Please don't stop at that! Your next step will be to get full information about the "Honor Bilt" Home you want to build.

Fill in our Information Blank RIGHT NOW!

Have us give you a definite proposition on the house you want to build. Then you will know to the penny just what your material will cost delivered to your station.

Fill in and mail the Information Blank today!

You can soon enjoy the comforts of a NEW "Honor Bilt" Home of your own.

## Please Fill this Out →

**Our Liberal Easy Payment Plan and Terms Described on Page 144**

**Our Guarantee Is Backed by Resources of One Hundred Million Dollars**

Which Modern Home do you like best?_____

_____Price, $_____

To what station would shipment be made?_____

_____State_____

Do you want to have an estimate of any of the following items for the house?

### CHECK ITEMS WANTED:

**HEATING**
- ☐ Hot Water
- ☐ Steam
- ☐ Warm Air Furnace
- ☐ Pipeless Warm Air Furnace
- ☐ To Burn Hard Coal or Soft Coal

**LIGHTING**
- ☐ Electric Fixtures

**PLUMBING**
- ☐ Bathroom Outfit Complete
  (Including Bathtub, Lavatory and Closet, Kitchen Sink, Range Boiler and Laundry Tub)
- ☐ Storm Doors and Windows
- ☐ Screen Doors and Window Screens

Prices in our Book of Modern Homes include lath for plaster.
Would you like us to figure on Goodwall Sheet Plaster, described on page 109, for use in place of lath and plaster?   ☐ Yes   ☐ No

About when do you expect to build?_____

## Your Name and Address Here →

Your Name_____

Postoffice_____

State_____

Rural Route_____ Box No._____

Street and No._____

# Sears, Roebuck and Co.
## Chicago~ Philadelphia~ Kansas City

# Your Next Step!

Please use the following blank lines to ask any questions you may have in mind that are not included in the Information Blank on the other side, or give us any other information you think we ought to know.

_____
_____
_____
_____
_____
_____
_____
_____
_____
_____
_____
_____
_____
_____
_____
_____
_____
_____
_____
_____
_____
_____

## EASY PAYMENT PLAN INFORMATION BLANK

If you are interested in our Easy Payment Plan (see page 144), please answer the following questions in addition to the ones on the other side of this page. We will then be glad to tell you just how our liberal terms will work out in your particular case.

1. Do you hold **legal title** to the lot or land on which you intend to build?_____ Frontage?_____ Depth?_____

2. How much did it cost? $_____ How much paid? $_____ Unpaid? $_____

3. Street paved?_____ Sidewalk?_____ Water?_____ Sewer?_____ Gas?_____ Electricity?_____

4. How much cash will you have to invest in the deal? $_____ How much can you pay each month? $_____

5. What is your occupation?_____.

6. If you plan to do any part or all of the work yourself, then tell us what work you can do?_____

_____
_____

## Sears, Roebuck and Co.—_The World's Largest Store_
### CHICAGO—PHILADELPHIA—KANSAS CITY

# Honor Bilt Modern Homes
# ...Order Blank...

**FOUR METHODS OF PAYMENT**

1. Cash With Order;
2. Pay as Building Goes Up;
3. Letter of Deposit;
4. Easy Payments.

---

IS THERE A FREIGHT AGENT at your railroad station?.............

HOW TO SEND MONEY. The best and safest ways to send money are by Postoffice Money Order, Express Money Order, Bank Draft or Check.

**DO NOT WRITE IN THIS SPACE**

DATE......................192..

NAME.........................................

POSTOFFICE....................................

RURAL ROUTE .............BOX No.........STATE...............

STREET AND No...................................

If You Want This Order Shipped to a Town Different From Your Postoffice Give Directions Here.

..........................................

..........................................
(Please state the road over which you wish shipment made.)

**State Total Amount of Money**
**═══ Sent With This Order ═══**

| DOLLARS | CENTS |
| --- | --- |
|  |  |

---

## 1. CASH WITH ORDER

Sears, Roebuck and Co.

In accordance with your guarantee of satisfaction or money returned, enclosed please find........................
(State whether Postoffice or Express Money Order, Draft, Check or Letter of Deposit.)

for $..........for your.........................
(State name of home wanted.)

"Honor Bilt" Modern Home No....................
(State number of home wanted.)

★Standard Built Home No....................
(State number of home wanted.)

described on page.............of your Book of "HONOR BILT" Modern Homes. This order is sent with the understanding that the price includes Lumber, Lath, Millwork, Sash Weights, Hardware, Nails, Paint, Building Paper, Eaves Trough, Down Spout, Roofing Material sufficient to complete the house in a workmanlike manner and according to your simple building plans which are furnished with the material.

★Price includes Lumber, Lath, Millwork, Sash Weights, Hardware, Nails, Paint, Tarred Felt, Roofing Material sufficient to complete the house in a workmanlike manner and according to your simple building plans which are furnished with the material.

**YOUR MONEY, INCLUDING ANY TRANSPORTATION CHARGES YOU PAID, WILL BE IMMEDIATELY RETURNED TO YOU FOR ANY GOODS NOT PERFECTLY SATISFACTORY.**

(Signed).........................................

Color of paint for outside body........................

Color of paint for outside trim........................
(See Page 20 for Colors.)

## 2. PAY FOR THE MATERIAL AS THE BUILDING GOES UP

If you are building by an arrangement with a bank or building and loan association, we will ship the material to you upon receipt of a letter from the bank or the building and loan association stating that the amount of our bill has been set aside for us and payments for the material will be remitted to us at certain periods as the work progresses.

## 3. PAY AFTER INSPECTING MATERIAL

If you would like to inspect material before sending us the money you can deposit the payment required with your bank or building and loan association and send us with your order this **letter of deposit**. This will give you a chance to inspect your material before we get a cent of your money.

### LETTER OF DEPOSIT

....................................192....

Postoffice.......................................    State..........................

Sears, Roebuck and Co.

.........................................has deposited with us the sum of $..............................
which has been set aside for you as a special fund, same to be paid to you upon delivery of the building material ordered of you, payments to be made at time material is delivered.

(SEAL)

Sears, Roebuck and Co. hereby agree that the material will be exactly as represented and entirely satisfactory to the depositor. The material is to be inspected upon its receipt and, if satisfactory, accepted by the depositor, who will then notify us to send Sears, Roebuck and Co. the money.

.........................................
(Name of Bank or Building and Loan Association.)

.........................................
(Signature of Officer and Title.)

# Sears, Roebuck and Co.—*The World's Largest Store*
## CHICAGO—PHILADELPHIA—KANSAS CITY

---

# The Greatest Building Proposition!

## 4. EASY PAYMENT PLAN

If you own a good, well located building lot free and clear from debt, and have some cash on hand, you can buy from Sears, Roebuck and Co. an "HONOR BILT" Modern Home, consisting of Lumber, Lath, Millwork, Sash Weights, Hardware, Nails, Paint, Building Paper, Eaves Trough, Down Spout and Roofing Material, Plumbing Goods, Heating Plant and Lighting Fixtures, or a STANDARD BUILT Home, on easy payment terms, and in some instances we will advance part of the cash for labor and material, such as brick, lime and cement, which we do not furnish.

### TO THOSE WHO ARE GOING TO DO THEIR OWN WORK:

Many of our customers build their own "HONOR BILT" homes, doing part or all of the work themselves, in which case they do not need to put in much cash, as we consider the value of their work the same as cash.

**A small payment each month,** as you would pay rent, makes you the owner of a good home in a few years. Interest charges, 6 per cent per year. Your interest becomes less each time you make a payment.

### THIS IS THE WAY OUR MONTHLY EASY PAYMENT PLAN WORKS OUT:

**Explanation:** The following is just an example to give you an idea of our LIBERAL EASY PAYMENT PLAN. It is not necessary that your lot cost exactly $500.00. It may have cost more. Likewise, it is not necessary that the labor and masonry material cost exactly $2,000.00. This may be more or less, depending upon the conditions in your locality. If you feel that you can comply with our few simple requirements, just **FILL IN THIS INFORMATION BLANK** and we will be glad to give you figures for your particular case **WITHOUT OBLIGATION** on your part.

Suppose you want to own a house to cost $4,500.00 complete, not including lot, and pay for it in future years.

|   |   |   |
|---|---|---|
| 1. | Cost of building lot............................................................ | $ 500.00 |
| 2. | Cost of building material and freight charges................................... | $2,500.00 |
| 3. | Cost of labor and masonry material, including any work you can do yourself............ | 2,000.00 |
|  | Cost of house (built complete)................................................ | 4,500.00 |
|  | Cost of lot and house completed.............................................. | $5,000.00 |

Here is what Sears, Roebuck and Co. will do in this case:

|   |   |   |
|---|---|---|
| 1. | Ship the material according to our monthly payment plan........................ | $2,500.00 |
| 2. | Advance you cash to help you pay for the construction of the building............... | 500.00 |
|  | Total................ | $3,000.00 |

This means that if you own a well located lot free from debt, and have sufficient cash, adding what it would cost for the work you can do yourself, if any, you can build a home with our assistance and pay us back a little each month, about equal to your rent.

### HERE IS WHAT WOULD BE REQUIRED OF YOU

**1.** Show clear title to building lot.

**2.** Send us with your order a payment equal to 2½ per cent of the amount of our material bill and cash we are to furnish. This is the additional charge we make to cover the expense of handling the transaction and compensate us for expenses we do not have in connection with our cash with order business.

**3.** Make small payment of one per cent (1%) or more, including interest each month. You may pay more than your regular monthly payment whenever you desire or even take up the loan in full before it is due.

## EASY PAYMENT PLAN INFORMATION BLANK

1. Which of our "HONOR BILT" Modern Homes or STANDARD BUILT Homes do you like best?......................

........................................................................Price...........................

2. Shall we include heating, plumbing, lighting or Goodwall sheet plaster? Check items wanted.

**HEATING**        **LIGHTING**

☐ Hot Water.     ☐ To burn Hard Coal     ☐ Electric Fixtures.     ☐ Goodwall Sheet Plaster.

☐ Steam.     ☐ or Soft Coal.     ☐ Plumbing

☐ Warm Air Furnace.

☐ Warm Air Pipeless Furnace.

3. Do you own the lot or land on which you intend to build?........Feet, frontage?...........Feet, depth?...........

4. How much did it cost? $...............How much paid? $...............How much unpaid? $...............

5. Street paved?..........Sidewalk?..........Water?..........Sewer?..........Gas?..........Electricity?..........

6. How much cash have you        How much can you
to put in the deal? $.............................................. pay each month? $...................

7. If you plan to do any part or all of the work yourself then tell us what work you can do...........................

8. About when do you expect to start building?.......................................................

Your Name......................................................9. What is your occupation?.....................

Postoffice....................................................State...........................

Rural Route..............Box No.............Street and No.......................................

## Sears, Roebuck and Co.—*The World's Largest Store*
### CHICAGO—PHILADELPHIA—KANSAS CITY

# Sears, Roebuck and Co's MILL WORK Products

The beautiful Dutch Colonial home shown to the left is our Priscilla, described on pages 34 and 35.

## The Millwork for "Honor Bilt" Modern Homes Is of the Highest Grade

See Pages 110 and 111 for additional information.

The illustrations on this page reveal the beautiful color effects obtained by the use of our various stains, enamels, varnishes and paints. See descriptions of houses in this book.

A Colonial front entrance of an "Honor Bilt" Modern Home.

French Doors. Finished with White Enamel 30—2165.

**SLAB PANEL DOOR SET**
Finished with Stain 30—2647 and Varnish 30—2719, with Enamel 30—2165 for the trim.

**VERTICAL PANEL DOOR SETS**
Stain 30—2646 and Varnish 30—2719.

**CLASSICAL "HONOR BILT" PERGOLAS**
Our Specifications for "Honor Bilt" Homes Give Complete Directions for Finishing.

**INTERIOR TWO-PANEL DOOR SETS**
Stain 30—2648 and Varnish 30—2719, with Enamel 30—2165 for the trim.

SEARS, ROEBUCK AND CO., CHICAGO-PHILADELPHIA

P598